METAL ION
ACTIVATION OF DIOXYGEN

METAL IONS IN BIOLOGY

EDITOR: **Thomas G. Spiro,**
Department of Chemistry
Princeton University, Princeton, New Jersey 08540

VOLUME 1 *Nucleic Acid—Metal Ion Interactions*
VOLUME 2 *Metal Ion Activation of Dioxygen*

In Preparation
Volume 3 Copper Proteins
Volume 4 Iron-Sulfur Proteins
Volume 5 Zinc Proteins

Metal Ion Activation of Dioxygen

Edited by

THOMAS G. SPIRO
Princeton University

A WILEY-INTERSCIENCE PUBLICATION

JOHN WILEY & SONS

New York   Chichester   Brisbane   Toronto

Copyright © 1980 by John Wiley & Sons, Inc.

All rights reserved. Published simultaneously in Canada.

Reproduction or translation of any part of this work beyond that permitted by Sections 107 or 108 of the 1976 United States Copyright Act without the permission of the copyright owner is unlawful. Requests for permission or further information should be addressed to the Permissions Department, John Wiley & Sons, Inc.

Library of Congress Catalog Card Number: 79-13808
ISBN 0-471-04398-2

Printed in the United States of America

10 9 8 7 6 5 4 3 2 1

Series Preface

Metal ions are essential to life as we know it. This fact has long been recognized, and the list of essential "trace elements" has grown steadily over the years, as has the list of biological functions in which metals are known to be involved. Only recently have we begun to understand the structural chemistry operating at the biological sites where metal ions are found. This has come about largely through the application of powerful physical and chemical structure probes, particularly X-ray crystallography, to purified metalloproteins. From such studies we have learned that nature has evolved highly sophisticated ways of controlling the relatively flexible stereochemistry of metal ions. In one case after another, the structure and reactivity of a metalloprotein active site has turned out to be different from anything previously encountered in simple compounds of the metals. Indeed, many a reasonable inference about active site structure, based on the known properties of metal complexes in solution, have turned out to be erroneous. These surprises have inspired inorganic chemists to expand their vision of metal ion reactivity. The biological studies have spurred much fruitful synthetic and mechanistic work in inorganic chemistry, aimed at elucidating the means whereby nature achieves its stereochemical ends. The terra incognita of the biochemical functions of metal ions has become familiar territory to an increasing number of inorganic chemists and biochemists, and several of the more imposing mountains have been scaled. Vast stretches remain uncharted, and the field is alive with a sense of both accomplishment and new opportunities.

The purpose of this series is to convey some of this excitement, as well as the emerging intellectual shape of the field, to a wide audience of nonspecialists. Individual volumes will cover topics that are current and exciting—the recently scaled mountains that are still under active exploration. The chapters are not intended to be exhaustive reviews of the subject matter. Rather, they are intended to be readable accounts of the insights and directions that are emerging in active new areas of research. Volumes will appear on an occasional basis as progress in the field dictates.

THOMAS G. SPIRO

Princeton, New Jersey

Preface

This second volume of the series focuses on how metal ions in biological systems activate molecular oxygen toward reduction of and/or insertion into organic molecules. This subject is currently under intensive investigation on a wide scientific front.

The first chapter, by James P. Collman, Kenneth S. Suslick, and Thomas R. Halbert, deals with the binding of oxygen to the heme group, a central theme in biological O_2 chemistry. The elegant work of the Collman group on the "picket fence" porphyrins, which has provided both structural and thermodynamic characterizations of O_2 binding to heme in a protein-free system, is described. In the second chapter, Minor J. Coon and Ronald E. White present an account of the important monooxygenase enzyme cytochrome P-450, with emphasis on the liver microsomal P-450 system, the elucidation of which owes much to the careful studies carried out in Coon's laboratory. In Chapter 3, John T. Groves discusses the varied mechanisms by which metal ion catalyze O_2 insertion that have been revealed by chemical studies. Our understanding of biological mechanisms must conform to these basic chemical principles. Groves' own work on iron catalysis provides much insight into the likely mechanisms for cytochrome P-450. In Chapter 4, John M. Wood discusses dioxygenases, the nonheme iron enzymes that insert both atoms of O_2 directly into aromatic substrates. His laboratory is responsible for many of the recent results on these enzymes. In Chapter 5, Bo G. Malmstrom presents a masterful and lucid account of cytochrome oxidase, the complicated and fascinating enzyme system that is responsible for the energy-coupled four-electron reduction of O_2 to H_2O in respiration. Finally in Chapter 6, James A. Fee gives a critical analysis of the much-discussed role by superoxide and the metallosuperoxide dismutases in biology.

These chapters do not exhaust the many themes of O_2 and metal ions in biology. They do give a sense of the breadth and fascination of this important subject.

T. G. SPIRO

Princeton, New Jersey
February 1980

Contents

1. O$_2$ Binding to Heme Proteins and Their Synthetic Analogs   1
   James P. Collman, Thomas R. Halpert, and Kenneth S. Suslick

2. Cytochrome P-450, A Versatile Catalyst in Monooxygenation Reactions   73
   Minor J. Coon and Ronald E. White

3. Mechanisms of Metal-Catalyzed Oxygen Insertion   125
   John T. Groves

4. Recent Progress on the Mechanism of Action of Dioxygenases   163
   J. M. Wood

5. Cytochrome c Oxidase   181
   Bo G. Malström

6. Superoxide, Superoxide Dismutases, and Oxygen Toxicity   209
   James A. Fee

   Index   239

METAL ION
ACTIVATION OF DIOXYGEN

CHAPTER **1**
O$_2$ Binding to Heme Proteins and Their Synthetic Analogs

JAMES P. COLLMAN

Department of Chemistry, Stanford University,
Stanford, California

THOMAS R. HALBERT

Corporate Research Laboratories,
Exxon Research and Engineering
Linden, New Jersey

KENNETH S. SUSLICK

Department of Chemistry, University of Illinois,
Champaign-Urbana, Illinois

CONTENTS

1 SYNTHETIC ANALOGS IN BIOINORGANIC CHEMISTRY, 3

2 STRUCTURE AND FUNCTION OF O_2-CARRYING HEMOPROTEINS, 5

2.1 Myoglobin, 7
2.2 Hemoglobin, 9
2.3 Nonmammalian Hemoproteins, 13

3 AXIAL COORDINATION TO IRON AND COBALT PORPHYRINS: GENERAL CONCEPTS, 14

3.1 Iron(II) Porphyrins, 14
3.2 Iron(III) Porphyrins, 22
3.3 Oxidation of Ferrous Porphyrins by O_2, 25
3.4 Cobalt Porphyrins, 28

4 DIOXYGEN BINDING TO IRON PORPHYRINS, 33

4.1 Low-Temperature Studies, 33
4.2 Kinetic Studies, 35
4.3 Polymer-Supported Systems, 39
4.4 Protected Pocket Porphyrins, 45
    4.4.1 Synthesis, 45
    4.4.2 Physical Properties of Iron Porphyrinate–O_2 Complexes, 47
        Electronic Spectra, 47
        Vibrational Spectra, 47
        Mossbauer Spectra and Magnetic Susceptibilities, 48
        Structural Data, 51
    4.4.3 Thermodynamics of O_2 Binding, 54
        Stability to Irreversible Oxidation, 54
        O_2 Binding to "R"-Type Hemoproteins and Analogs, 55
        O_2 Binding to "T"-State Hb and Analogs, 60
        Cooperativity in O_2 Binding to Iron Porphyrinates, 62

REFERENCES, 66

1 SYNTHETIC ANALOGS IN BIOINORGANIC CHEMISTRY

This chapter deals with the bioinorganic chemistry of myoglobin and hemoglobin. In particular, the work discussed provides a successful example of the synthetic analog approach to the study of metalloproteins. This technique relies on the design and synthesis of low molecular weight compounds which are structurally and functionally similar to the active site of metalloproteins. The study and characterization of such synthetic analogs may allow a more detailed knowledge of the chemistry of the active site than can be easily obtained from the protein itself. The chronology of synthetic analog studies generally follows this sequence:

1. Isolation and purification of the metalloprotein.
2. Measurement of physical properties and preliminary characterization of the protein's active site.
3. Synthesis and isolation of analog complexes.
4. Physical and structural characterization of the synthetic analog complexes.
5. Comparison between the protein and the analogs and revelation of new structure–function relationships.

One may note in this protocol that molecular modeling occurs only after a great deal is already known about the protein. This is dictated by the very nature of the synthetic analog methodology: it is difficult to design rationally and synthesize active site model compounds unless one already has considerable information about the active site. This however, does not diminish the utility of the approach. Small-molecule analogs can provide a detailed understanding of the structure and functioning of the active site which cannot be obtained (or at least not without extreme difficulty) by studying the metalloprotein itself. More importantly, new insights as to the role of the protein in manipulating the reactivity of the metal may be provided only by removal of those influences. Finally, the ability to mimic enzymatic reactions has potentially important technological consequences. A fundamental postulate of bioinorganic chemistry is the capability of relatively small metal complexes to emulate the chemistry of metallobiomolecules. Closely tied to this assumption is the belief that the function and mechanism of biological systems can be understood in terms of fundamental chemical principles.

A useful classification of metal-containing biomolecules is shown in Figure 1. The O_2-binding proteins myoglobin and hemoglobin are part of the large group of transport and storage metalloproteins. Myoglobin and

Abbreviations

ATP	adenosine triphosphate
BM	Bohr magneton
bzacen	N,N'-ethylenebis(benzoylacetoneiminato)
CoMb	cobalt-substituted myoglobin
CoHb	cobalt-substituted hemoglobin
Ct–Np	distance from center of porphyrin hole to a pyrrole nitrogen
DPIXDME	deuteroporphyrin IX dimethyl ester
DMA	dimethylacetamide
DMF	dimethylformamide
DPG	2,3-diphosphoglycerate
EPR	electron paramagnetic resonance
EtOH	ethanol
EXAFS	extended X-ray absorption fine structure
Hb	hemoglobin
HMPA	hexamethylphosphoramide
Im	imidazole
Mb	myoglobin
1-MeIm	1-methylimidazole
2-MeIm	2-methylimidazole
1,2-Me$_2$Im	1,2-dimethylimidazole
meso-PIXDME	mesoporphyrin IX dimethyl ester
N$_{Im}$	imidazole nitrogen
NMR	nuclear magnetic resonance
N$_p$	pyrrole nitrogen in a porphyrin
OEP	octaethylporphyrin
Piv$_3$(4CImP)Por	*meso*-tri(α,α,α-*o*-pivalamidophenyl)-β-*o*-5)*N*-imidazoyl)butyramidophenylporphyrin
Piv$_3$(5CImP)Por	*meso*-tri(α,α,α-*o*-pivalamidophenyl)-β-*o*-5(*N*-imidazoyl)valeramidophenylporphyrin
P$_N$	mean plane of porphyrin nitrogens
PPIX	protoporphyrin IX
PPIXDME	protoporphyrin IX dimethyl ester
py	pyridine
THF	tetrahydrofuran

TPP	*meso*-tetraphenylporphyrin
TpivPP	*meso*-α,α,α,α-tetra-*o*-pivalamidophenylporphyrin
T(*p*-OCH$_3$)pp	*meso*-tetra(*p*-methoxyphenyl)porphyrin
1-TrIm	1-tritylimidazole
Val	valine

hemoglobin are not enzymes. They do not catalyze reactions of dioxygen; they simply transport and store it. Even so, these proteins are among the most studied of all molecules, biological or not. Hemoglobin in particular is a biochemical book of world records: the first protein crystallized, the first associated with a specific physiological function, one of the first to have its molecular weight measured, the first eukaryotic protein to have its messenger RNA isolated and characterized, the first to be synthesized in vitro without whole cells, and the first associated with a genetic disease (1). Synthetic analog studies of these proteins have provided for productive interactions between biochemists and inorganic chemists, which probably is also a first.

2 STRUCTURE AND FUNCTION OF O$_2$-CARRYING HEMOPROTEINS

In all creatures of any size, transport of O$_2$ must be facilitated. Only when the organism's surface area is very large compared to its internal volume can simple diffusion provide sufficient O$_2$ to the interior cells. Similarly the tissues must be able to store O$_2$ for use under conditions of sudden demand. In a few invertebrates these tasks are performed by a copper-containing protein (hemocyanin, responsible for the blue blood of lobsters) or by a nonporphyrin iron-containing protein (hemerythrin, responsible for the pink blood of marine worms). In most organisms however it is hemoglobin and myoglobin which serve these functions. These respiratory proteins are found in all vertebrates, in some worms and insects, and even in the root nodules of leguminous plants. Myoglobin and hemoglobin are structurally similar to the cytochromes (which provide electron transport and are linked to oxidative phosphorylation), the monooxygenases (which hydroxylate drugs and hormones; in general, catalyzing the reaction —C—H → —C—OH), the hydroperoxidases (which detoxify peroxides and oxidize various substrates), and the terminal oxidases (which provide the electron sink for oxidative catabolism by catalyzing

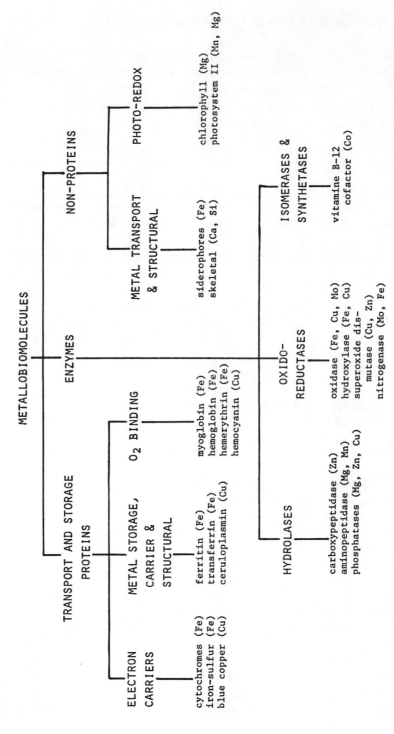

Figure 1. Classification of metal-containing biomolecules.

the reduction of O_2). These hemoproteins are thought to have evolved perhaps a billion years ago and thus are one of the oldest classes of biomolecules.

2.1 Myoglobin

Myoglobin (Mb), a relatively simple protein found primarily in vertebrate muscles, consists of two components: the heme cofactor (an Fe(II) protoporphyrinate IX complex) and a protein globin of about 150 amino acid residues. The tertiary structure of globins consists of eight α-helical regions labeled A through H (from the anion to carboxyl ends) and six short nonhelical segments (labeled NA, AB, CD, EF, FG, and GH), together forming a basket for the heme. The heme is held in place by a large number of nonpolar and hydrogen-bonding interactions at its periphery and by a covalent bond between the iron atom at the N_ϵ of the "proximal" imidazole of F8 histidine. The tertiary structures of a large number of myoglobins have been determined by X-ray or neutron diffraction methods. The structure of sperm whale metmyoglobin [i.e., containing Fe(III)] proposed by Kendrew et al. (2) was in fact the first protein structure solved. Since that time, structures have been presented for seal (3), tuna (4), and lamprey (5) myoglobins as well. In all cases the general tertiary structure is the same and is shown schematically in Figure 2. In addition, the myoglobins of more than 20 vertebrate species have been sequenced (1). Some 80 of the 150 residues are invariant: 20 are responsible for the heme–globin contacts, 20 are involved in interresidue hydrogen bonds. The geometry of the iron site has been roughly established by these structures. In the deoxy ferrous myoglobins, the iron forms a five-coordinate complex (to the four porphyrin nitrogens and the proximal imidazole) and lies out of the porphyrin plane toward the proximal side. Upon oxygenation or other ligand binding, the iron atom moves into the plane forming a six-coordinate complex. Further details of this process are discussed later.

The O_2 binding characteristics of myoglobins are uncomplicated. The O_2 binding at the single heme is a simple equilibrium following the usual mass action law:

$$K = \frac{[Mb \cdot O_2]}{[Mb][O_2]}$$

This is often expressed in terms of the fraction of protein oxygenated,

$$Y = \frac{[Mb \cdot O_2]}{[Mb] + [Mb \cdot O_2]}$$

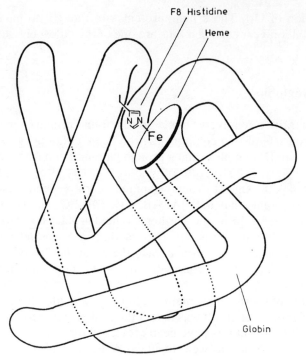

Figure 2. Tertiary structure of myoglobin.

so that

$$\frac{Y}{1-Y} = K[O_2]$$

If the concentration of O_2 is expressed in terms of pressure, then $K^{-1} = P_{1/2}$, the O_2 pressure at which half-saturation occurs. In general, for myoglobins at room temperature $P_{1/2}$ is roughly 1 torr. There are few complicating features to the interaction of myoglobin with O_2: there are no effects due to the concentrations of protein, salt, or buffer and only small changes in $P_{1/2}$ with pH changes (the Bohr effect) (6). Similarly the kinetics of O_2 binding are straightforward and follow a simple bimolecular rate expression with $k \sim 2 \times 10^7$ M^{-1} sec^{-1} at room temperature, which is near diffusion control considering the limited access created by the small opening of the protein's heme pocket (7).

Myoglobin can reversibly bind a number of other ligands. These include (6) nitric oxide, carbon monoxide, and various isonitriles (RNC), which are bound respectively $\sim 10^5$, ~ 30, and ~ 0.1 as strongly as O_2. Much work has been reported on the ligand-binding properties of the ferric ion in metmyoglobin. In general, metmyoglobin prefers anionic ligands, such as N_3^- or CN^-, and no longer interacts with O_2 or CO. These studies have revealed some information about the nature of the binding pocket, but we will not discuss them further in this chapter.

The biological function of myoglobin is primarily O_2 storage in muscle tissue. In this way myoglobin acts as an intermediate between blood O_2 transport (hemoglobin) and tissue oxidative phosphorylation (terminating with cytochrome oxidase). The concentration of myoglobin within the muscles depends upon the physiological need for O_2 storage. Skeletal muscles of aquatic mammals, for example, have very large amounts of myoglobin, as does heart muscle. Myoglobin concentrations can also be increased when needed, as in high-altitude acclimatization. Thus myoglobin functions as an O_2 buffer to counteract fluctuations in the organism's demand or supply of O_2.

2.2 Hemoglobin

Hemoglobin (Hb) is the O_2-transporting protein of the blood. As a generic term "hemoglobin" has been used for a wide range of hemoproteins found in mammals, other chordates, invertebrates, and even some plants and protozoa (which of course have no blood at all). Mammalian Hb is the prototype of multisubunit (oligomeric) proteins and consists of a dimer of dimers: two subunits designated α, which have approximately 140 amino acid residues, and two β subunits with approximately 145 residues. Each of these subunits contains one heme molecule capable of binding one molecule of O_2. Extensive X-ray structural studies have been made of hemoglobin in various states of ligation and have been recently reviewed (1). The tertiary structures of the subunits are always similar in general terms to those of myoglobin. Both the tertiary structures of the individual subunits and the quaternary structure (i.e., the intersubunit contacts and relative packing) of the tetramer show significant and specific changes upon ligation at the iron atoms. The various quaternary structures observed have generally been classified into two categories (1): the "R" and "T" states, corresponding to those found in oxygenated and deoxygenated molecules, respectively.

This variation in structures is reflected in the O_2 binding properties of mammalian Hb. The O_2 affinities of these two structural classes are quite different; the $P_{1/2}$ for the R state is slightly smaller (i.e., higher affinity) than that of the isolated, monomeric subunits or simple myoglobins, whereas the $P_{1/2}$ for the T state is much larger (i.e., much lower affinity). If we plot the log of the oxy/deoxy ratio ($Y/1 - Y$) against the log of the O_2 pressure (P_{O_2}) (a Hill plot), then for O_2 binding sites which operate independently, as in monomeric myoglobin, a linear correlation is expected with unit slope.

Hemoglobin however has four O_2 binding sites, and these sites do *not* operate independently. Instead, the tetramer functions in a coordinated manner, wherein the binding of O_2 to one subunit increases the O_2 affinities of the others. This is made obvious in the Hill plot shown in Figure 3. When completely deoxygenated Hb (T conformation) is exposed to low O_2 pressures, noncooperative, low-affinity binding occurs, as reflected by a unit slope region in the Hill plot. As the O_2 pressure is increased, a slope greater than unity is observed, indicative of interaction between sites (i.e., cooperativity) in O_2 binding. As the O_2 pressure is further increased, Hb again displays noninteractive O_2 binding, but with high affinity, leading to the fully oxygenated R conformation. The slope of the intermediate region is the Hill coefficient n and aproaches 3 in native Hb. Since it is often the case that data can only be easily gathered within this

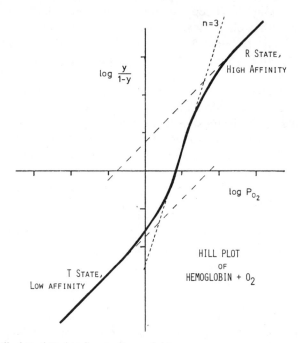

Figure 3. Hill plot of O_2 binding to hemoglobin.

intermediate range, the O_2 binding to Hb is often approximated by the Hill equation:

$$\frac{Y}{1-Y} = \left(\frac{P_{O_2}}{P_{1/2}}\right)^n$$

It should be emphasized that n has no direct physical meaning; a better measure of the interheme interaction is the ratio of limiting $P_{1/2}$ values at low versus high O_2 pressures (6). This ratio varies from 10 to 1000, depending on conditions.

The mechanism of Hb cooperativity remains a controversial subject. Two distinct approaches have been taken: a purely mathematical modeling of the observed O_2 binding characteristics, and a stereochemical, structural explanation. The complexity of the cooperative equilibria involved is readily apparent. Each heme can be in one of two states (O_2 bound or not), each subunit can be in one of two tertiary forms (oxy or deoxy), and the quaternary structure can be either R or T. In one limit these choices may be considered completely independently (so that *all* combinations of heme coordination, subunit tertiary structure, and quaternary structure are taken into account), giving rise to $2^4 \times 2^4 \times 2$, or 512 states, each with a different relative concentration and a different O_2 affinity! Mathematical models utilizing a variety of simplifying assumptions have been proposed to reduce the parameters needed to fit the observed O_2 binding data (6). The observed data however do not provide sufficient information to distinguish among any number of such models.

The stereochemical model of Hb cooperativity proposed by Perutz attempts to describe the structural events which occur in a Hb tetramer upon oxygenation and the manner in which these changes affect the O_2 affinity. The Perutz mechanism of Hb cooperativity has been reviewed elsewhere in detail (1) and is discussed later in this chapter. The stereochemical mechanism has been shown to lead to a mathematical description consistent with the observed O_2 binding (8).

The cooperative interactions which occur among the individual subunits of a Hb molecule are termed homotropic and are essential to efficient O_2 transport. The high O_2 affinity of hemoglobin at high O_2 pressures guarantees full saturation at the lungs, whereas its low affinity at low O_2 pressures ensures complete transfer of O_2 from Hb to Mb in the tissues. Figure 4 shows graphically this complementary nature of Hb transport to storage of O_2 by Mb. It is also physiologically important to be able to moderate the efficiency of this transport as needed so as to give large amounts of O_2 to some organs at some times and small amounts to others. This is accomplished by heterotropic interactions between Hb and other mole-

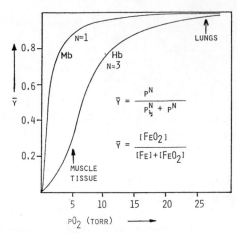

Figure 4. O_2 Binding to hemoglobin and myoglobin.

cules or ions. Such "effectors" fall into three categories (8): (a) competitive; (b) tertiary; and (c) quaternary. Competitive effectors are simply other ligands which will also bind to the ferrous porphyrinate. Thus in the presence of CO, which competitively binds to deoxy-Hb, O_2 binding is diminished and the apparent $P_{1/2}$ for O_2 is increased. Tertiary effectors are those molecules or ions which stabilize one subunit tertiary form over the other. For example, an intrasubunit ion pair ("salt bridge") between the protonated imidazole of β146 histidine and the carboxylate of β94 aspartic acid exists only in the deoxy tertiary conformation; thus increased H^+ concentration (i.e., lowered pH) enhances the stability of this salt bridge by increased protonation of the imidazole and hence stabilizes the deoxy tertiary conformation relative to the oxy conformation (9). This is the Bohr effect, the lowering of O_2 affinity with lowering of pH (named after Christian Bohr (the father of Niels Bohr) who discovered the effect in 1904). Quaternary effectors are those molecules or ions which stabilize one quaternary form over the other. Phosphates such as 2,3-diphosphoglycerate (DPG), adenosine triphosphate (ATP), or even just PO_4^{3-} affect O_2 binding in this way. In the deoxy, T quaternary state there is a cavity between the two β subunits (9) which contains a number of positively charged residues and is capable of binding phosphates or other anions quite tightly with a number of salt bridges. In the oxy, R quaternary state however this crevice is closed and no longer can easily bind phosphate. Thus phosphates, by binding selectively to the T quaternary conformation, stabilize the deoxy state relative to the oxy state and decrease the apparent O_2 affinity of hemoglobin.

The physiological importance of these effectors is obvious, since they allow for biochemical feedback between the O_2 transport and utilization systems. The organic phosphates are especially significant (10) because they are part of the oxidative phosphorylation process, which depends on O_2 transport by Hb. An illustrative example of the interdependence of structure, physiological function, and phosphate effectors is found in fetal hemoglobin. Fetal Hb is a tetramer consisting of two α subunits and two γ subunits, which replace the adult β chains. These γ subunits have one of the cationic residues of the phosphate binding site replaced with a neutral one thereby reducing the phosphate affinity of fetal Hb. In the absence of organic phosphates, the O_2 binding of fetal and adult Hb is identical; under physiological conditions however fetal Hb has a higher affinity than adult Hb, which is essential to placental O_2 transport. This difference is ascribed to the lowered affinity of fetal Hb for organic phosphates: the same level of DPG will not lower the O_2 affinity of fetal Hb as much as that of adult Hb (10).

2.3 Nonmammalian Hemoproteins

O_2-Carrying hemoproteins are not limited to mammals, having been found in most nonmammalian vertebrates, many invertebrates, and even a few plants and protozoa. Formally, "myoglobins" are *noncirculating* O_2-binding hemoproteins found most commonly in muscle tissues; "hemoglobins" are O_2-binding hemoproteins found in *circulatory cells*; and "erythrocruorins" are extracellular O_2-binding hemoproteins found in blood or other circulating fluids (11). Many authors however simply refer to all reversible O_2-binding proteins in vertebrates as "hemoglobins."

In nearly all cases these hemoproteins are composed of subunits which are structurally similar to myoglobin: one heme in a subunit weighing 16,000 to 20,000 daltons. The heme is nearly always protoporphyrin IX, except in a few polychaete annelid worms which contain chlorocruoroporphyrin in which the 2-vinyl group of protoporphyrin has been replaced by a formyl group (notwithstanding its name, this porphyrin has no chlorine; it simply has a green color). The number of subunits that make up the hemoproteins is variable. Some proteins are monomeric (e.g., leghemoglobin of soy bean nodules) (12), others are dimeric (e.g., those found in the trachael cells of *Gastrophilus intestinales*, a stomach worm parasite), most are tetrameric (e.g., vertebrate hemoglobins), a few are octameric (e.g., the perienteric fluid of *Ascaris lumbricoides*, a nematode) (14), and some are polymeric, containing hundreds of subunits (e.g., the erythrocruorins of various annelid worms and mollusks and human sickle cell hemoglobin (15, 16).

The vast differences in the physiological demands of these various organisms suggest that large variations should exist in the functional properties (i.e., O_2 binding) of their hemoglobins. This is observed in the $P_{1/2}$ of O_2 binding to hemoproteins at room temperature which vary from ~0.002 torr for the nematode *Ascaris* (14), to ~1 torr for mammalian Mb, to ~1000 torr for trout blood at pH 6 (6). This 500,000-fold difference (which corresponds to an 8 kcal/mole difference in the ΔG of O_2 binding) clearly demonstrates the impact which the protein globin may have on the chemical reactivity of the iron porphyrin complex. One of the primary reasons for making synthetic analogs of these O_2-binding hemoproteins is to understand the ways in which different proteins so dramatically moderate the chemical behavior of the same prosthetic iron porphyrinate complex.

3 AXIAL COORDINATION TO IRON AND COBALT PORPHYRINS: GENERAL CONCEPTS

The axial coordination chemistry of a simple metalloporphyrin, in the absence of special constraints imposed by a protein, is governed by the electronic nature of the metal-chelate complex and the stereochemical requirements of the relatively rigid planar tetradentate macrocycle. Before one can begin to understand the subtle factors involved in the control of reversible dioxygen binding to heme proteins and their synthetic analogs, a general knowledge of the axial ligation chemistry of simple metalloporphyrins is necessary. Key information needed includes thermodynamic data for binding of nondioxygen ligands having varying electronic and steric demands, detailed structural parameters, and spectroscopic data gathered by utilizing a wide range of techniques. Fortunately for the researcher in hemoprotein chemistry, a massive body of such information has been published, and the extensive literature has been reviewed in depth (17). More recent reviews, with emphasis on biologically relevant metalloporphyrins, are also available (1, 18–20).

It is our intent in this section to present only an overview of information that bears directly on the question of dioxygen binding in hemoproteins and synthetic analogs. Therefore we limit the discussion to iron and cobalt systems.

3.1 Iron(II) Porphyrins

Iron in the 2+ oxidation state is the metal ion of the most obvious importance in hemoprotein chemistry. Iron(II) has six *d* electrons, and therefore a four-coordinate iron porphyrin has two vacant coordination sites. When two bases bind, they are constrained by the macrocycle to lie in axial positions, trans to one another across the ring. The primary axial coordination processes for unconstrained ferrous porphyrins are illustrated in Figure 5; the two states of biological interest are the five-coordinate state with B = an imidazole, which is an analog of the iron site in deoxy-Hb and deoxy-Mb, and the corresponding six-coordinate O_2 complex, which serves as an analog to oxy-Hb and oxy-Mb.

Although isolation of simple porphyrins in either of these states has only been achieved recently under carefully controlled conditions, in the early 1960s Williams (21) and Hoard (22, 23) realized that spin state–structure relationships in model compounds could lead to important insights into the function of hemoproteins. On the basis of the limited structural information then available, Hoard (22) deduced that in a strain-free porphyrin, the radius of the central hole (distance from the center of the porphyrin to a pyrrole N, the Ct–Np distance) is 2.01 Å. He then reasoned that high-spin Fe(II) has too large a covalent radius to fit into this core, and therefore that high-spin five-coordinate Fe(II) would lie substantially out of the porphyrin plane toward the single axial ligand. Furthermore the smaller low-spin six-coordinate ferrous ion of the oxyprotein was predicted to lie in the plane. It was noted that the expected motion of the ferrous iron could provide a starting point for a cooperativity mechanism (23). These ideas were incorporated as a key part of the mechanism of cooperativity suggested by Perutz (24) in which motion of the iron into the porphyrin plane upon oxygenation is responsible for the changes in

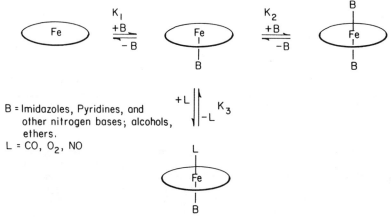

Figure 5. Axial coordination of iron(II) porphyrins.

protein conformation that lead to increased O_2 affinity. At the time these concepts were developed, there were very little actual data on ferrous porphyrins. The mass of information gathered since has supported the structural predictions, although some of the rationale has been modified, as we shall see.

At present, well characterized examples of all the ferrous porphyrin–ligand combinations illustrated in Figure 5 are known. The stereochemistries and spin states of the various species can be rationalized to a reasonable extent on the basis of simple crystal field theory as demonstrated in Figure 6. The equilibrium constants governing the formation of many of the complexes in solution are also known, and representative data are presented in Table 1. We now briefly examine each of the major combinations in turn.

Square planar unligated ferrous porphyrins can be prepared by reduction of ferric porphyrins in noncoordinating solvents provided O_2 is rigorously excluded (26, 28, 29). Magnetic moments of ~4.4 Bohr magnetons (BM) are typically observed, suggesting the intermediate spin state $S = 1$. This spin state results from the relatively low energy of the d_{z^2} orbital in the absence of axial ligands; the orbital occupation scheme illustrated in Figure 6 has been suggested based on extensive Mössbauer and NMR studies (30, 31). The X-ray crystal structure (30) of square α,β,γ,δ-tetraphenylporphyrinatoiron(II), Fe(TPP), shows the iron precisely centered in the porphyrin core with an Fe–Np bond length of 1.972(4) Å. In order to accommodate this short bond length, the porphyrin is ruffled in a manner described by Hoard (22) as a circular standing wave with crests at

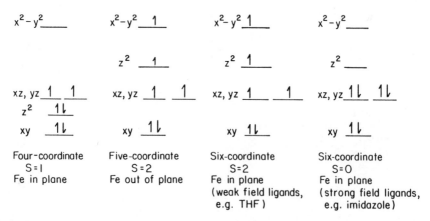

Figure 6. Structure and spin-state properties of ferrous porphyrins.

Table 1 Equilibrium Constants for Binding of Axial Bases to Iron(II) Porphyrins (25–27)[a]

Porphyrin	Base	K_1 (M^{-1})	K_2 (M^{-1})	K_3(CO) (M^{-1})[b]
DeuteroPIXDME	Im	4.5×10^3	6.8×10^4	4.8×10^8
DeuteroPIXDME	2-MeIm	1.3×10^4		2.5×10^6
DeuteroPIXDME	py	$>1.6 \times 10^3$	$>8.1 \times 10^4$	
DeuteroPIXDME	4-CN-py	4.7×10^3	7.1×10^4	5.6×10^7
TPP	Im	8.8×10^3	7.9×10^4	
TPP	py	1.5×10^3	1.9×10^4	
DeuteroPIXDME	CO	5×10^4	21	
DeuteroPIXDME	THF	5.2	0	2.8×10^6
DeuteroPIXDME	Dioxane	2.5	0	3.8×10^6
DeuteroPIXDME	DMF	3.4	0	4.3×10^6
DeuteroPIXDME	EtOH	1.5	0	8.0×10^6
DeuteroPIXDME	H$_2$O	~0.1	0	~5.0×10^6

[a] All in benzene at 25°C.
[b] [CO] in benzene = 6.7×10^{-3} M/atm.

one pair of oppositely situated methine carbon atoms and troughs of equal magnitude at the other pair.

Five-coordinate high-spin ($S = 2$) ferrous porphyrins are more difficult to prepare, and the problem clearly stems from the fact that $K_2 > K_1$ for simple aromatic nitrogen base ligands, as can be seen from Table 1. The low-spin six-coordinate d^6 complexes are favored thermodynamically by crystal field stabilization energy. With $K_2 > K_1$, the desired five-coordinate species cannot be present in any large proportion in solution. Collman and Reed reasoned that with the sterically hindered axial base 2-methylimidazole (2-MeIm), steric interaction between the methyl group and the porphyrin ring would prevent the iron from moving into the porphyrin plane and thus prevent six-coordination (28). The desired five-coordinate complex was isolated as the ethanol solvate, Fe(TPP)(2-MeIm)·C$_2$H$_5$OH. The magnetic moment of 5.2 BM (25°C) is close to both the spin-only value for an $S = 2$ ion (4.9 BM) and the measured moment for deoxy-Hb (5.1 BM) (24). More recently, the high-spin five-coordinate 2-MeIm adduct of the Fe(II) "picket fence" porphyrin (2-MeIm)Fe(TpivPP) has also been prepared, and crystal structures of both deoxy-Mb models have been determined (33, 34).

The key structural parameters of the coordination groups are shown in Figure 7. The high-spin iron lies 0.40 to 0.42 Å out of the mean plane of the porphyrin nitrogens (P$_N$). In the TPP complex the displacement of

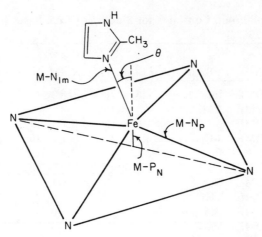

Figure 7. Geometry of five-coordinate metalloporphyrins. FeTpivPP(2-MeIm): $M–P_N$ = 0.40 Å; $M–N_p$ = 2.072; $M–N_{Im}$ = 2.095; θ = 11°. FeTPP(2-MeIm)·EtOH: $M–P_N$ = 0.42 Å; $M–N_p$ = 2.086; $M–N_{Im}$ = 2.161; θ = 9.6°.

the iron from the mean plane of the whole porphyrin (Ct) is actually 0.13 Å greater (0.55 Å) due to extreme doming of the porphyrin believed to result from crystal packing forces. In the TpivPP complex the doming is only 0.03 Å, typical of most five-coordinate metalloporphyrins. In both cases the Fe-to-2-MeIm bonds are longer than the Fe-to-imidazole bonds observed for low-spin six-coordinate complexes [e.g., 2.01 Å in Fe(TPP)(1-MeIm)$_2$ (18)]; this is expected based on occupation of the d_{z^2} orbital in the high-spin iron. The longer Fe–N_{Im} distance in the TPP complex may be attributed to steric effects caused by the more nearly eclipsing conformation which the imidazole plane adopts with respect to the Fe–porphyrin nitrogen bonds (Fe–N_p). The degree of eclipsing is expressed by the angle ϕ between a projection of the imidazole plane and an Fe–N_p bond, as shown in Figure 8. Although ϕ in the model compounds is again largely controlled by crystal packing forces, the observation of the importance of ϕ suggests the intriguing possibility that the globin in the proteins may control the out-of-plane distance of the iron at least in part by controlling this angle.

Several lines of evidence point to the conclusion that the out-of-plane displacement in the models is close to the thermodynamically optimal one for high-spin *five-coordinate* ferrous porphyrins and does not result from steric distortion caused by the methyl group in 2-MeIm. First, any such distortion would be expected to reduce K_1 for 2-MeIm binding. As Rougee and Brault have shown however, K_1 for 2-MeIm with Fe(TPP) is actually slightly greater than for imidazole itself (Table 1). Second, the crystal structures show no short nonbonding interactions between the porphyrins and the axial ligands. Finally, the displacement of the high-spin Mn(II) from the porphyrin plane in Mn(TPP)(1-MeIm) is 0.56 Å (35); this value is actually somewhat larger than that for the Fe(TPP)(2-MeIm), despite the fact that 1-MeIm has no steric constraints.

Reed has recently pointed out the relevance of these results to the Hoard–Perutz cooperativity mechanism (18) which initially included a protein-induced "tension" on the proximal histidine believed to stretch the Fe–N_{Im} bond and pull the metal further out of the plane in deoxy-Hb. The displacement of the iron from the heme plane in human deoxy-Hb A is best estimated at ~0.6 Å (36), while the Fe–Ct distance in Fe(TPP)(2-MeIm) is 0.55 Å. Thus any "strain" induced by the protein on the out-of-plane position of the iron must be slight. Such an argument must be viewed with some caution however because, as we have seen, the Fe–Ct distance includes 0.13 Å doming, and the Fe–P_N distance in the less severely domed Fe(TpivPP)(2-MeIm) is only 0.40 Å. Furthermore the accuracy with which the Fe–Ct distance in the protein is known is probably not better than ±10%. Somewhat more convincing arguments against "tension" have resulted from experiments employing a wide variety of other physical techniques. Such work includes resonance Raman studies carried out by Spiro and his associates (37) on the 2-MeIm adduct of Fe(II) mesoporphyrin IX and deoxy-Hb, EXAFS studies by Eisenberg and collaborators (38), and NMR studies by Goff and LaMar (39).

Although there are at present no other simple five-coordinate ferrous porphyrins with aromatic nitrogen bases for which crystal structures are available, two other general synthetic approaches have been used to enforce five-coordination. One approach involves covalently attaching a single axial base to a porphyrin; the other, blocking the approach of the sixth ligand through construction of a "cap" over one face of the por-

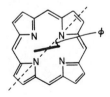

Figure 8. Projection of axial imidazole plane on porphyrin plane.

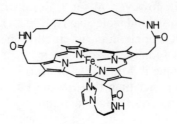

Figure 9. Battersby's "tailbase" porphyrins.

phyrin. Traylor and co-workers (40–42) prepared ferrous mesoporphyrins and pyrroporphyrins modified by covalent attachment of imidazole or pyridine to one of the propionic acid side chains; however the five-coordinate state was characterized only by solution visible spectra. Momenteau (43) covalently linked histidine to a propionic acid side chain of Fe(II) deuteroporphyrin, while Battersby constructed the porphyrin shown in Figure 9 (44). Again characterization was limited to visible spectra. Halbert (45, 46) prepared tetraphenylporphyrins with imidazole attached via an alkyl chain to the ortho position of a single phenyl ring. The high-spin five-coordinate complexes were characterized by magnetic susceptibility, elemental analysis, NMR, Mössbauer studies, visible spectroscopy, and magnetic circular dichroism. One difficulty with such "tailbase" porphyrins is that the $K_2 > K_1$ problem becomes significant at low temperatures. Visible spectra and Mössbauer and NMR studies demonstrate that at temperatures much below room temperature the "tail" porphyrins dimerize to give mixed six-coordinate/four-coordinate dimers (43, 45).

The "capping" approach has been studied by Baldwin (47), who prepared the elegant porphyrin shown in Figure 10. The pyromellitic acid-

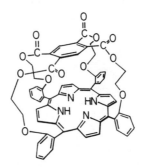

Figure 10. Baldwin's "capped" porphyrin.

derived "cap" effectively blocks one side of the porphyrin. Again the five-coordinate ferrous pyridine and N–MeIm adducts have been characterized by solution spectra only. A number of other attempts have been made to enforce five-coordination using less rigid "straps" across one face of the porphyrin (48–51); a typical example is Baldwin's "strapped" porphyrin shown in Figure 11. In the ferrous complexes reported, the straps are apparently flexible enough to permit six-coordination (48).

With weakly coordinating ligands such as THF (25–27) as well as with CO (52) and OH^- (53), the existence of five-coordinate ferrous complexes has been inferred from solution visible spectra. The CO complex of Fe(II)TPP has also been shown to be diamagnetic by NMR; a full structure would be of interest to determine the position of the iron with respect to the porphyrin plane.

As previously mentioned, in the absence of special constraints nitrogen bases invariably give diamagnetic six-coordinate ferrous porphyrin complexes. Such complexes are typified by (piperidine)$_2$Fe(TPP), the full X-ray crystal structure of which shows the iron to be exactly in the porphyrin plane, with a M–N$_p$ distance of 2.004(3) Å (54). This is in keeping with Hoard and Williams' hypothesis that the low-spin ferrous ion, with unoccupied d_{x-y^2} and d_{z^2} orbitals (see Fig. 5) should exhibit a smaller effective radius and fit nicely into the porphyrin core. Recently however this rationale involving the size of high-spin versus low-spin Fe(II) has been questioned on several grounds. First, structural studies of compounds such as $[Fe(TPP)(OH_2)_2]^+$ (55), $Sn(Cl)_2(TPP)$ (56), and $Mo(TPP)(O_2)_2$ (57) exhibit an in-plane metal with M–N$_p$ distances substantially larger than 2.01 Å, thus suggesting that the porphyrin core radius is more variable than originally thought and could accommodate a high-spin ferrous ion. Second, theoretical calculations have suggested that minimization of non-

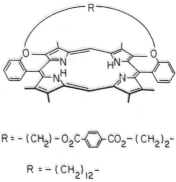

$R = -(CH_2)-O_2C-\langle \rangle-CO_2-(CH_2)_2-$

$R = -(CH_2)_{12}-$

Figure 11. Baldwin's "strapped" porphyrin.

bonded repulsion between the axial ligand and the porphyrin nitrogen orbitals is *at least* as important as the size of the ion in determining the out-of-plane displacement of the metal (58, 59). Finally, and possibly most dramatically, Reed and Scheidt have succeeded in crystallizing the high-spin, six-coordinate ferrous porphyrin $(THF)_2Fe(TPP)$ and have shown that the iron essentially is centered in the porphyrin plane (60). The high-spin state has been verified by magnetic susceptibility and Mössbauer measurements and is surely the result of the weak field nature of the tetrahydrofuran ligands, which give rise to a crystal field splitting qualitatively shown in Figure 5. The crystal structure shows a $M-N_p$ distance of 2.057 Å, substantially larger than the presumed optimal core size of 2.01 Å. In short, it now seems that for six-coordinate porphyrins in which the repulsive interactions of the two axial ligands balance, the metal will lie in the plane, with $M-N_p$ as large as about 2.1 Å. Furthermore, in five-coordinate porphyrins, the out-of-plane distance will depend on a delicate balance involving minimization of repulsive interactions between the ligand and the porphyrin, maximization of $M-N_p$ bonding by centering of the metal, and minimization of strain energy involved in deforming the porphyrin core from its optimal radius of 2.01 Å.

The only other general coordination types for ferrous porphyrins (besides O_2 complexes) are those involving CO and NO as axial ligands. We will not discuss nitrosyls, except to say that crystal structures of both five- and six-coordinate complexes have been carried out, primarily by Scheidt, who has recently reviewed the area (61). Carbon monoxide complexes of simple porphyrins with aromatic nitrogen bases as the sixth axial ligand are low spin, with a linear Fe–C–O unit perpendicular to the porphyrin plane. Representative structural parameters are those for (py)(CO)FeTPP with $Fe-N_p$ of 2.02 Å and the iron displaced a slight 0.02 Å from the porphyrin nitrogen plane (P_N) toward the CO (62). Though the linear Fe–C–O unit is not surprising, it is of interest because of the observation of a "bent" or "tilted" CO group in carbon monoxy-Hb and -Mb. This distortion is believed to result from steric interactions with the distal Val-E11-CH_3 and the His-E7 imidazole groups. Such steric interaction has been proposed to lower the CO affinity and thus protect the heme from endogenous CO poisoning (63).

3.2 Iron(III) Porphyrins

As previously mentioned (Section 2.1), ferric porphyrins do not bind O_2 and might thus seem of less direct relevance to a review on O_2 binding. Because of the susceptibility of ferrous porphyrins toward oxidation (see Section 3.3) however, a great deal of work on simple ferric porphyrins, met-Hbs, and met-Mbs (ferric proteins) has been carried out, and many of the arguments regarding cooperativity revolve about spin state–structure relationships for the ferric state (1).

The primary axial coordination chemistry of simple ferric porphyrins is illustrated in Figure 12 (64–66). The high-spin ($S = \frac{5}{2}$) five-coordinate complexes 1 and 2 with anionic axial ligands are well known, and a number of X-ray crystal structures have been reported. Typical parameters are Fe–N_p = 2.065 and Fe–Ct = 0.45 Å (67). As was the case with the five-coordinate ferrous porphyrins, the out-of-plane displacement must result from steric interactions between the ligand and the porphyrin as much as it results from the size of the high-spin ferric ion; this conclusion is supported by the recent isolation of high-spin six-coordinate species involving weak field ligands, in which the iron is located exactly in the porphyrin plane (65) (e.g., 6, B = H_2O, Fe–N_p = 2.04 Å, Fe–Ct = 0). NMR studies have also demonstrated the existence of high-spin six-coordinate species in solution (66).

The dimeric μ-oxoporphyrins 3 are readily formed in the presence of hydroxyl ion. Although each iron is five-coordinate and presumably high spin $S = \frac{5}{2}$, the magnetic moment is typically about 2.4 BM at room temperature. This low moment is the result of a strong antiferromagnetic coupling, which has been studied in detail by Mössbauer spectroscopy

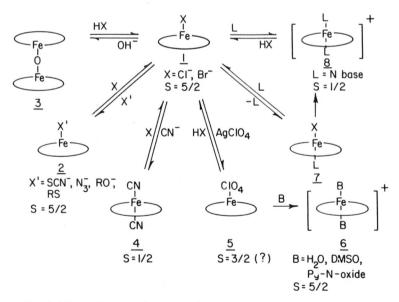

Figure 12. Axial coordination chemistry of iron(III) porphyrins.

(68) and magnetic susceptibility measurements (69) over a wide temperature range. As with other high-spin five-coordinate porphyrins, the iron is substantially displaced from the porphyrin plane (Fe–Ct ~0.50 Å) and the M–N_p distances are large (~2.09 Å) (70).

While there are no known unligated ferric porphyrins analogous to the intermediate-spin ($S = 1$) four-coordinate ferrous porphyrins, intermediate-spin ($S = \frac{3}{2}$) ferric porphyrins appear to have been realized in the cases of the unusual five-coordinate perchlorate complexes **5** with octaethylporphyrin (71) and TPP (65). Observed magnetic moments are somewhat high (between 4.5 and 5.0 BM at room temperature), but Mössbauer studies support an $S = \frac{3}{2}$ formulation (71). Structural data on the TPP complex (65) include a small out-of-plane displacement (Fe–Ct = 0.30 Å) relative to high-spin ferric complexes, and distinctly short M–N_p distance (1.99 Å).

In the absence of special constraints ferric porphyrins, like their ferrous counterparts, yield only six-coordinate low-spin ($S = \frac{1}{2}$ for ferric) complexes **8** with aromatic nitrogen bases. Although the determination of equilibrium constants for axial ligation in ferric porphyrins has received only limited attention, the binding of aromatic nitrogen bases has been examined by several authors. In general it has not been possible to observe the intermediate in which a single imidazole is bound; some typical equilibrium constants for the overall reaction $Fe^{III}porphyrin(X) + 2L \rightleftharpoons [Fe^{III}porphyrin(L)_2]^+ X^-$ are given in Table 2. A significant point to emerge from this work is that the binding constants are extremely sensitive to solvent polarity (which is not surprising in light of the ionic nature of the species involved).

The problems in preparing mixed-ligand six-coordinate complexes like **7** with simple porphyrins, graphically demonstrated by the equilibrium

Table 2 Equilibrium Constants for Aromatic Nitrogen Base Binding to Fe(III) Porphyrins

Porphyrin	Base	K	Solvent	Reference
$Fe^{III}TPPS^a$	Imidazole	1.7×10^6	H_2O (pH 6)	72
$Fe^{III}DPIXDME^b$	Imidazole	1.4×10^7	$CHCl_3$	43
$Fe^{III}TPP$	Imidazole	4.8×10^5	CH_2Cl_2	73, 74
$Fe^{III}TPP$	Imidazole	$\sim 5 \times 10^4$	Benzene	73
$Fe^{III}TPPS$	Histidine	1.8×10^5	H_2O (pH 6)	72
$Fe^{III}TPPS$	Pyridine	1.5×10^3	H_2O (pH 5)	72

[a] TPPS = Tetra(p-sulfonato)phenylporphyrin.
[b] DPIXDME = Deuteroporphyrin IX dimethyl ester.

studies above, have made it difficult to model directly the Fe(III) site in the various forms of met-Hb and met-Mb, in which the protein supplies a single axial histidylimidazole and the sixth ligand is varied. Extensively studied met-Hb and -Mb states include those in which the sixth ligand is F^- (predominantly high spin, $S = \frac{5}{2}$), H_2O, OCN^-, OH^-, SCN^-, N_3^-, NO_2^- (thermal equilibrium between $S = \frac{1}{2}$ and $S = \frac{5}{2}$), and CN^- (low spin, $S = \frac{1}{2}$). The mixed-spin proteins have been of particular interest because of the relationship between the quaternary state of the appropriate met-Hb forms and the position of the spin equilibrium (1). Nevertheless the only model compound available which resembles any of these forms is $(py)(N_3^-)Fe^{III}(TPP)$, which is a low-spin ($S = \frac{1}{2}$) complex with the metal in the porphyrin plane and with $Fe-N_p$ = 1.99 Å (67). The most logical approach to better models is through porphyrins with a single covalently attached axial imidazole, but the few efforts in this direction have been complicated by a highly favorable dimerization equilibrium similar to that discussed in Section 3.1 for ferrous "tail" porphyrins (43, 45).

3.3 Oxidation of Ferrous Porphyrins by O_2

In contrast to the reversible oxygenation of Hb and Mb, simple ferrous porphyrins are rapidly and irreversibly oxidized by oxygen. An understanding of the mechanism of this oxidation is crucial to the design of kinetically stable model iron porphyrin dioxygen complexes. In addition, such knowledge clarifies the role of the protein in stabilizing the ferrous state. Early studies on the kinetics of oxidation of ferrous porphyrins in nonaqueous media were carried out by Caughey (75–77) and Wang (78). With high concentrations of base and low pressures of oxygen (<0.2 atm), the rate law is of the form shown in Eq. (1) below. With low base concentrations and higher O_2 pressures (0.5 to 1 atm), the reaction becomes first order in B_2Fe^{II} and the O_2 dependence becomes more complex. The final product of oxidation was found to be the previously described μ-oxo dimer:

$$\frac{-d[Fe^{II}]}{dt} = k\frac{[B_2Fe^{II}]^2[O_2]}{[B]^2} \quad (1)$$

In parallel experiments on the autoxidation of $FeCl_2$ in EtOH and MeOH, Hammond and Wu (79) also found the rate to be second order in ferrous ion and first order O_2. One equivalent of O_2 was found to oxidize four equivalents of Fe(II) to Fe(III). Significantly, when benzoin was added to the reaction mixture, it was oxidized to benzil without affecting the rate of oxidation or the form of the rate law. This observation led

Hammond and Wu to postulate the intermediacy of a "ferryl" species, $Fe(IV)=O$, which would be expected to be a potent oxidant.

Based on the above observations, the general mechanistic scheme outlined in Figure 13 can be proposed (75, 79). Strong additional support for this scheme has come recently from NMR studies carried out by LaMar and co-workers (80), who observed that a μ-peroxo intermediate similar to **13** can be stabilized at −50°C in toluene during the oxidation of base-free solutions of Fe(II) tetra(*m*-tolyl)porphyrin. This work also points to the existence of an oxidation pathway parallel to that in Figure 13 but starting with the four-coordinate porphyrin species **9**. The only intermediate shown for which there is at present no direct chemical evidence is the ferryl species **14**, which is analogous to "Compound II" in horseradish peroxidase (81). Further work on isolation or characterization of such potent oxidants is clearly warranted.

Although we have not included possible acid-catalyzed oxidation pathways in Figure 13, there is reason to believe such alternate pathways

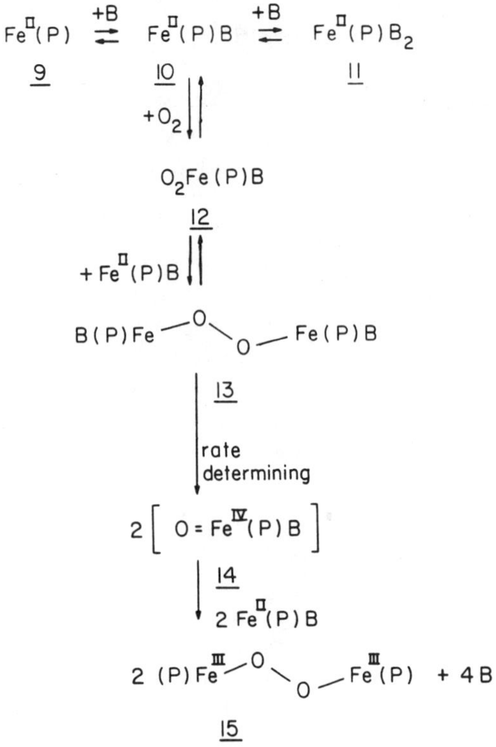

$$Fe^{II}(P) \underset{9}{\overset{+B}{\rightleftarrows}} Fe^{II}(P)B \underset{10}{\overset{+B}{\rightleftarrows}} Fe^{II}(P)B_2 \quad 11$$

$$+O_2 \updownarrow$$

$$O_2Fe(P)B$$
$$12$$

$$+Fe^{II}(P)B \updownarrow$$

$$B(P)Fe{-}O{\diagdown}_{O}{-}Fe(P)B$$
$$13$$

rate determining

$$2\left[O=Fe^{IV}(P)B\right]$$
$$14$$

$$\downarrow 2\,Fe^{II}(P)B$$

$$2\,(P)Fe^{III}{-}O{\diagdown}_{O}{-}Fe^{III}(P) + 4B$$
$$15$$

Figure 13. Oxidation of ferrous porphyrins; (P) represents the porphyrin ligand.

exist. For example, in aqueous solution the overall rate of oxidation of Fe(II) salts is increased by addition of acid (82, 83). In contrast however, Hammond and Wu (79) found no effect in nonaqueous media. Apparently an alternate oxidation pathway, similar to the speculative pathway shown in Eqs. (2)–(4), competes with the pathway of Figure 13:

$$Fe-O_2 + H^+ \rightarrow Fe^{III} + HO_2 \tag{2}$$

$$Fe^{II} + HO_2 + H^+ \rightarrow Fe^{III} + H_2O_2 \tag{3}$$

$$2Fe^{II} + H_2O_2 + 2H^+ \rightarrow 2Fe^{III} + 2H_2O \tag{4}$$

The acid-catalyzed pathway dominates only in aqueous acid or under special conditions where formation of a μ-peroxo intermediate is hindered. Once again, further work in this area is needed.

Consideration of the oxidation pathways outlined above leads to two general observations regarding the role of the protein globin in preventing rapid and irreversible autoxidation of Mb-O_2 and Hb-O_2. First, by isolating the Fe(II) sites from one another, the protein prevents formation of the μ-peroxo intermediate **13** and effectively blocks oxidation from occurring via the primary mechanism of Figure 13. Second, the globin limits access of protons to the oxygen binding site through judicious arrangement of amino acid residues with nonpolar or basic side chains (e.g., valine E11 and histidine E7), thus preventing oxidation.

Understanding the mechanistic pathways by which iron–dioxygen complexes irreversibly decompose also permits the synthetic chemist to identify several ways of stabilizing such complexes in the absence of the protecting globin. For example, attachment of the porphyrin to a rigid synthetic polymer or construction of a porphyrin with suitable steric constraint around the iron could prevent formation of the μ-peroxo dimer intermediate **13** and thus block oxidation. In addition, anything which lowers the relative concentration of the five-coordinate species **10** will inhibit oxidation by inhibiting formation of the μ-peroxo intermediate **13**. High O_2 pressures and high base concentrations accomplish this by driving the equilibria towards the O_2 complex **12** and the six-coordinate complex **11**, respectively. Low temperatures have a similar effect in that they tip the equilibria toward the thermodynamically more stable six-coordinate complex **11** and oxygen complex **12**. Of course low temperatures also slow the rate-determining μ-peroxo cleavage directly, as LaMar has demonstrated (80).

In one form or another, all of the above observations have been exploited during the past several years by investigators seeking to prepare and study dioxygen complexes of iron porphyrins. In Section 4, we examine details of the many approaches and discuss their relative merits in more detail.

3.4 Cobalt Porphyrins

In all natural hemoglobin and myoglobin proteins, the prosthetic group is an iron heme. However the replacement of the natural iron heme in Hb and Mb with a synthetic Co(II) heme has been accomplished (84–88). Cobalt-substituted hemoglobin and myoglobin (CoHb and CoMb) are functional artificial O_2-carrying proteins, although their O_2 affinities are 10 to 100 times less than those of their natural iron counterparts. Furthermore, CoHb exhibits cooperativity, although to a lesser extent than Hb. In the same way that structural and spectroscopic properties of simple iron porphyrin complexes have been employed to help unravel the mysteries of O_2 binding in Hb and Mb, studies on simple cobalt porphyrins have been employed to explain the properties of CoMb and CoHb. Furthermore, comparisons between the iron and cobalt systems have proved useful in testing theories of cooperativity (1, 84–90). A brief review of the axial coordination chemistry of Co(II) porphyrins follows; the chemistry of Co(I) and Co(III) porphyrins is less well developed and of little interest in the context of this chapter.

Figure 14 summarizes the primary axial coordination processes observed for Co(II) porphyrins. Cobalt(II) is d^7, and thus Co(II) porphyrins could in theory be found in any of several spin states. In practice, all Co(II) porphyrins are found to be low spin, $S = \frac{1}{2}$. The parent four-coordinate complex is structurally similar to the corresponding four-coordinate iron complex; the cobalt is in the porphyrin plane, and the short Co–N_p distance (1.949 Å for CoTPP) gives rise to an S_4 "ruffling" of the porphyrin framework (92). EPR spectra, which have proved especially useful in characterizing Co(II) porphyrins, demonstrate that the single unpaired electron resides primarily in the d_{z^2} orbital (89).

Four-coordinate Co(II) porphyrins react with a wide variety of Lewis bases, affording the five-coordinate complexes **17**. In contrast to the Fe(II) porphyrins however, the five-coordinate species **17** are always thermodynamically favored over the six-coordinate species **18** in solution; this fact is illustrated in Table 3, which contains typical equilibrium constants. The small values of K_2 are commonly rationalized by pointing out that the d_{z^2} orbital which must participate in the axial bonding is partially occupied in d^7 low-spin Co(II). In addition, no ligand field stabilization energy is gained in going from the five-coordinate to the six-coordinate state, since the Co(II) is low spin in both cases.

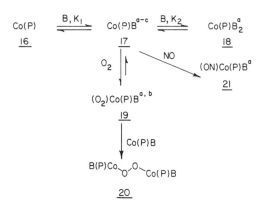

^{a} B = Aromatic nitrogen bases, piperidine
^{b} B = tetrahydrothiophene, HMPA, DMF, DMA, CNMe, PR$_3$, PF$_3$, P(OR)$_3$
^{c} B = Co, NO

Figure 14. Axial coordination chemistry of cobalt(II) porphyrins (89–91).

The structures of the five-coordinate nitrogen base adducts of Co(II) are typified by (1-MeIm)Co(TPP), with Co–N$_p$ = 1.977 (3) Å and Co–N$_{Im}$ = 2.157(3) Å (96). The cobalt lies 0.14 Å out of the mean plane of the porphyrin, a small displacement in comparison with 0.42 Å in (2-MeIm)Fe(TPP). This smaller displacement is expected, since the $d_{x^2-y^2}$ oribital is unoccupied in the Co(II) case and the effective radius of the cobalt is small enough for the ion to be accommodated more readily by

Table 3 Equilibrium Constants for Binding of Axial Bases to Co(II) Porphyrins^a

Porphyrin	Base	K_1	K_2	Reference
T(p-OCH$_3$)PP^b	Piperidine	2400	0.8^c	94, 95
T(p-OCH$_3$)PP	3,4-Dimethylpyridine	1100	0.1^c	94, 95
PPIXDME^d	Pyridine	1110	0^e	89
PPIXDME	1-MeIm	3300	0^e	89
PPIXDME	Tetrahydrothiophene	23	0^e	89

^{a} In toluene, units l/mole.
^{b} Tetra(p-methoxyphenyl)porphyrin, 25°.
^{c} Calculated from ΔH and ΔS in references 94 and 95.
^{d} Protoporphyrin IX dimethyl ester.
^{e} No six-coordinate complex observed.

the porphyrin core. In fact, the observed 0.14 Å displacement probably results primarily from the steric interactions between the axial ligand and the porphyrin as previously discussed for the iron case. Interestingly, when the more sterically demanding 1,2-Me$_2$Im ligand is used, the displacement is increased to 0.16 Å and the Co–N$_{Im}$ bond is stretched an additional 0.06 Å to 2.216(2) Å (97). In addition, the Co–N$_{Im}$ bond is tipped by about 7° from a normal to the porphyrin plane. Of the other bases which give five-coordinate Co(II) porphyrin complexes, only the nitrosyl complex (NO)Co(TPP) has been studied by X-ray crystallography (98). Complexes of the other bases shown in Figure 14 have been characterized only by solution-spectroscopic techniques (91).

The six-coordinate Co(II) porphyrin complexes **18** can in certain cases be isolated by utilizing a large excess of axial base. Crystal structures are available for (Pip)$_2$Co(TPP) (99) and (3-methylpyridine)$_2$Co(OEP) (100). The cobalt is in the porphyrin plane as expected, with relatively short Co–N$_p$ bonds (1.98 to 1.99 Å). The axial Co–N bonds are quite long (2.436 and 2.386 Å, respectively). This is again expected because of the partial occupation of the d_{z^2} orbital.

The ability of Co(II) complexed with nonporphyrin ligands to bind O$_2$ was recognized long before comparable O$_2$ complexes of cobalt porphyrins were discovered. In 1852 Fremy reported what is probably the first synthetic O$_2$ complex, $[(NH_3)_{10}Co_2O_2]^{4+}$ (101). In a recent review McClendon and Martell list 23 characterized mononuclear Co–O$_2$ complexes, and 26 binuclear Co–O$_2$–Co complexes (102). Representative compounds from both classes have been extensively studied by such techniques as X-ray crystallography, Raman and infrared spectroscopy, electron paramagnetic resonance, and X-ray photoelectron spectroscopy. In the mononuclear complexes the O$_2$ binds in a bent end-on configuration and is often written as Co^{III}O$_2^{-1}$. The most accurate crystal structure to date is that determined by Schaefer and co-workers (103) for a Schiff base cobalt complex, Co(t-Bsalten)(1-benzimidazole)(O$_2$). The Co–O–O bond angle in this complex is 117.5°, and the O–O bond length is 1.27(1) Å, very close to the bond length of 1.28(2) Å in ionic superoxide. The binuclear complexes are described as bridging μ-peroxo complexes, with the cobalt atoms formally in the 3+ oxidation state again. Typically, Co–O–O bond angles are from 112 to 120° while O–O bond lengths vary from 1.31 to 1.47 Å depending on the nature of the other ligands on cobalt.

With most nonporphyrin ligands, the binuclear complexes are more stable, being readily formed at room temperature. The mononuclear complexes can be obtained however by use of low temperatures, high dilution, sterically hindered ligands, or nonaqueous solvents (102). Cobalt porphyrins in contrast show little tendency to form binuclear O$_2$ complexes.

The only evidence for formation of such a complex is the slow disappearance of the EPR signal (95) in samples containing a mixture of a mononuclear O_2 complex (**19** in Fig. 14) and a five-coordinate base adduct (**17** in Fig. 14), and the formation of a diamagnetic complex between O_2 and a cobalt containing cofacial, diporphyrin reported by Chang (104).

Mononuclear cobalt porphyrin O_2 complexes are readily formed at low temperature from the five-coordinate base adducts, although in most cases the equilibrium binding constants are small at room temperature (89, 90, 95, 104–109). A recent exception is the Co(II) "picket fence" porphyrin which is almost fully oxygenated under 1 atm O_2 at 20°C in toluene. $P_{1/2}$ values for 1 : 1 O_2 binding to selected Co(II) porphyrins are reproduced in Table 4. It is clear from this table that the choice of porphyrin ligand and axial base has a dramatic effect on the O_2 affinity. It has been proposed (90) that the greater affinity of the "picket fence" compounds for O_2 is the result of solvation effects; "flat" porphyrins like Co(Tp-OCH$_3$PP) and Co(PPIXDME) are better solvated by toluene in their deoxy forms, whereas the "picket fence" tends to level solvation effects between the deoxy and oxy forms (90). The way in which the axial base controls oxygen affinity has been a matter of substantial debate (84, 89, 108, 109). All of the data on bases other than imidazoles have been obtained with simple porphyrins which bind O_2 effectively only at low temperatures or high O_2 pressures. The difficulties of the experiments as well as the small temperature ranges available have led to wide variations in calculated ΔH^0 and ΔS^0 values (in contrast, the measured $P_{1/2}$ data vary much less at any given temperature). Differing sets of thermodynamic data have been used to contend that π-acceptor character (94), π-donor character (84), or σ-donor character alone (89) is responsible for the relative strength of O_2 binding with axial nitrogen bases. Although the questions of this nature are not yet resolved in detail, it is clear that a reasonable correlation exists between σ-donor strength of the axial base and O_2 affinity; Drago's

Table 4 $P_{1/2}$ Values for 1 : 1 O_2 Binding to Cobalt Porphyrins in Toluene

Porphyrin	Base	T (°C)	$P_{1/2}(O_2)$ (torr)	Reference
Co(Tp-OCH$_3$PP)	N-MeIm	25	15,500[a]	95
Co(PPIXDME)	N-MeIm	25	17,800[a]	105
			11,000[a]	89
Co(TpivPP)	N-MeIm	25	140	90
Co(PPIXDME)	N-MeIm	−31	178	89
Co(PPIXDME)	Pyridine	−30	5,500	89
Co(PPIXDME)	Tetrahydrothiophene	−35	25,300	89

[a] Calculated from data in the reference.

recent treatment of ΔH^0 data obtained at high O_2 pressures and low temperatures in terms of his E and C method (89) has been fairly successful.

Although the data in Table 4 all pertain to toluene solutions, the effect of solvents with higher dielectric constants has been studied in several cobalt–porphyrin–O_2 binding systems (84, 90, 95, 104). With the simple porphyrins, polar solvents lead to increased O_2 binding constants. However with the "picket fence" porphyrin, solvent polarity has little effect. Once again this is most likely the result of differential solvation of the deoxy and oxy forms of the simple porphyrins, although the direct effect of high dielectric constant on stabilizing the charge separation present in CoO_2 complexes may also be significant.

At present no X-ray crystal structures are available for cobalt–porphyrin–O_2 complexes. However IR (63) and EPR (84, 104, 110) studies suggest a close correspondence with the 1:1 cobalt–O_2 complexes involving nonporphyrin ligands. The O_2 is surely bound in the "bent end-on" fashion, and the O–O bond length will probably fall in the range of 1.25 to 1.35 Å. On the basis of EPR (84) and ESCA (111) studies, the unpaired electron in the cobalt–O_2 complexes is generally assigned to an

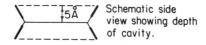
Schematic side view showing depth of cavity.

Figure 15. Baldwin's iron(II) tetraaza macrocycle, **22**.

oxygen π orbital. However on the basis of the EPR cobalt hyperfine coupling constants, an alternative "spin-pairing" model has been proposed (110) in which the unpaired electron in the Co d_{z^2} orbital couples with an unpaired electron in a dioxygen π-antibonding orbital. This model leads to the conclusion that the actual charge transferred to O_2 in the cobalt–porphyrin complexes is about 0.5 electron.

4 DIOXYGEN BINDING TO IRON PORPHYRINS

In Section 3.3 the mechanism of autoxidation of ferrous porphyrins was discussed, and several possible approaches to stabilizing dioxygen complexes of Fe(II) were pointed out. The principal approaches which have proven successful to date are low-temperature stabilization, attachment of the porphyrin to a rigid polymer, and construction of porphyrin ligands with steric constraints that prevent bimolecular oxidation. In addition, kinetic studies of O_2 binding have been carried out by a number of techniques which avoid complications due to oxidation. In the following sections we discuss the results of each of these varying approaches.

4.1 Low-Temperature Studies

The relative kinetic stability of iron–dioxygen complexes at low temperature in dilute nonaqueous solutions was first recognized in 1973 by Baldwin (112) and Traylor (40) and their collaborators. Baldwin utilized visible spectrophotometry to show that the Fe(II) macrocycle **22** in Figure 15 would reversibly bind O_2 in toluene containing 1% pyridine at −85°C. The 1:1 stoichiometry was verified by manometric O_2 uptake, but the complex oxidized irreversibly above −50°C. Baldwin suggested that the stability of this complex at low temperature is also due in part to the steric bulk of the phenyl rings, which might slow bimolecular reaction (see Section 4.4). In light of more recent work (*vide infra*) on unhindered systems, it now appears that low temperature is the more important factor in stabilizing this system.

Traylor's initial work (40) reported the reversible binding of O_2 by the Fe(II) porphyrin **23** in Figure 16, which has a covalently attached imidazole. Characterization was based on visible spectral changes in CH_2Cl_2 solution below −45°C. In addition to stabilization by low temperature, a "neighboring group effect" due to the appended based was proposed to lend stability. Later experiments (113) with the heme **23** led to an estimated $P_{1/2}(O_2)$ of 0.2 torr at −45°C in CH_2Cl_2. Spectral studies on the

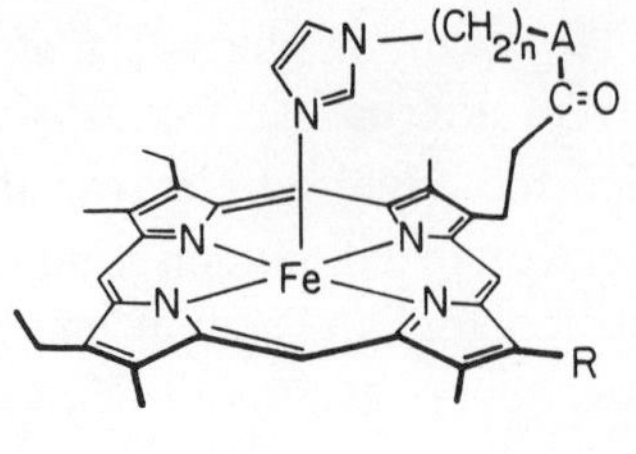

Figure 16. Porphyrins with covalently attached bases prepared by Traylor et al.

23: n = 3	R = H	A = NH
24: n = 3	R = CH$_2$CH$_2$CONH(CH$_2$)$_3$Im	A = NH
25: n = 2	R = CH$_2$CH$_2$CO$_2$CH$_3$	A = O
26: n = 3	R = CH$_2$CH$_2$CO$_2$CH$_3$	A = NH
27: n = 4	R = CH$_2$CH$_2$CO$_2$CH$_3$	A = NH

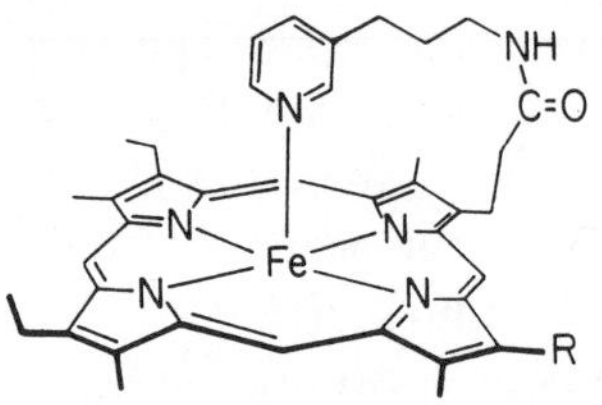

28: R = CH$_2$CH$_2$CO$_2$H
29: R = CH$_2$CH$_2$CO$_2$Me
30: R = H

pyridine heme **28** showed rapid oxidation at $-45°$ in CH$_2$Cl$_2$, perhaps due to the free acid group in the heme. The stated conclusion that axial pyridine does not support oxygenation now seems unjustified in light of later reports of reversible oxygenation of iron porphyrins with pyridine axial bases reported by Collman (114) and Basolo (115).

While the relative stability of the above O$_2$ complexes was ascribed at least in part to steric protection or neighboring group effects, the efficiency of low temperature *alone* in stabilizing nonaqueous solutions of O$_2$ complexes of simple porphyrins was dramatically pointed out in four papers published simultaneously in 1974 (115–118). Basolo and co-workers (115) investigated the reactions of Fe^{II}TPP(pyridine)$_2$, Fe^{II}TPP(1-MeIm)$_2$, and Fe^{II}TPP(piperidine)$_2$ with O$_2$ in dry CH$_2$Cl$_2$ at $-78°$. In all three cases they observed reversible oxygenation by visible spectrophotometry and, with the pyridine adduct, they determined by manometric techniques that 0.96 ± 0.1 mole of O$_2$ was taken up per mole of porphyrin. Experiments in toluene and diethyl ether also demonstrated qualitatively that the $P_{1/2}(O_2)$ values are smaller in solvents of high dielectric constant.

Baldwin and co-workers (116) found that Fe^{II}*meso*-PIXDME(1-MeIm)$_2$ will reversibly bind O$_2$ at $-50°C$ in CH$_2$Cl$_2$. They noted that spectral changes were identical to those observed by Traylor for porphyrins with appended imidazole, and verified the uptake of 1 mole O$_2$/Fe manometrically. They also pointed out that the use of free imidazole itself as the axial base leads to irreversible oxidation. Brinigar and Chang (117) studied the reversible oxygenation of Fe^{II}PPIXDME in DMF containing a 100-fold excess of 1-BuIm at $-45°C$. Visible spectral changes were used to estimate $P_{1/2}(O_2)$ of 80 torr, compared to an estimated $P_{1/2}(O_2)$ of less than 1 torr for the porphyrin **25** in Figure 16, which has two covalently attached imidazole ligands. The difference has been ascribed to a more favorable orientation of the imidazole enforced by the attaching chain in **25**. Different degrees of competition for the second axial site (O$_2$ versus excess free base or an appended base) would appear to make such a provocative conclusion rather tenuous. Finally, Wagner and Kassner found (118) that Fe^{II}PPIX in *n*-butanol containing a large excess ($>10^4$-fold) of 2-MeIm or *t*-butylamine could be reversibly oxygenated at $-80°C$. Visible spectra as well as Mössbauer spectra were presented as evidence. These authors also noted that while the deoxygenated solutions appeared to contain five-coordinate Fe(II) at room temperature, at $-80°C$ six-coordination prevails.

In summary, it is clear that simple ferrous porphyrins at low concentration (10^{-5} to 10^{-4} M) are capable of binding O$_2$ with minimal oxidation in the presence of a variety of axial bases, provided the temperature is below about $-40°C$. Use of dipolar aprotic solvents appears to improve oxidative stability. Alkyl imidazole, pyridine, and simple amine-type ligands, as well as DMF, are all suitable axial bases, though the imidazoles appear qualitatively to be better. Covalent attachment of the base seems to afford little in the way of added stability and may actually complicate evaluation of results at low temperature because of the dimerization equilibrium (2 five-coordinate → 1 six-coordinate + 1 four-coordinate) discussed in Section 3.1.

4.2 Kinetic Studies

It may come as a surprise to some that the widely used "rapid mixing" technique for the study of reaction kinetics on the time scale 0.01 to 1 sec was first developed and applied by Hartridge and Roughton (119) in 1923, in a study of the rate of dissociation of O$_2$ from oxyhemoglobin. They used rapid mixing of a solution of oxy-Hb with aqueous sodium dithionite (Na$_2$S$_2$O$_4$) in a flow cell. Because the rate of dithionite reaction with free O$_2$ is substantially faster than the rate of dissociation of oxy-

Hb, the kinetics of dissociation to deoxy-Hb could be determined spectrophotometrically.

Since these pioneering experiments, a number of other techniques have been applied to the study of the kinetics of O_2 binding to heme proteins. Temperature-jump experiments, in which an electrical discharge is used to heat the sample solution rapidly and thus disrupt the equilibrium, have been used in conjunction with visible spectrophotometry to obtain O_2 "on-rates" (k') and "off-rates" (k) for several myoglobins and hemoglobins (120–122). Flash photolysis methods for studying O_2 "on-rates" have been developed (123, 124) based on the extremely high quantum yields ($\phi \sim 1$) for photodissociation of CO from carbonmonoxyhemoproteins (125, 126). In these experiments, CO is flashed off the carbonmonoxyprotein in the presence of a relative excess of O_2, and the O_2 association is followed spectrophotometrically. Finally, several improvements on Hartridge and Roughton's initial experiments have been utilized, including stop–flow reaction of deoxyprotein solutions with O_2-containing solutions to obtain O_2 on-rates (124, 127) and stop–flow reaction of oxyprotein solutions with dithionite solutions to obtain O_2 off-rates (127).

The results of a number of experiments utilizing the above techniques are presented in Table 5. For Mbs, isolated α and β chains of Hb, and leghemoglobin, the oxygen association kinetics are always observed to follow a simple second-order rate law, while first-order kinetics are observed for O_2 dissociation. The measured rate constants are similar for all myoglobins and for the isolated Hb chains, while the abnormally high O_2 affinity of leghemoglobin is seen to derive primarily from an exceptionally *large on-rate*. The kinetics of binding to the tetrameric Hb molecule are much more complex. Ilgenfritz and Schuster (122) have quantitatively fit the temperature-jump relaxation kinetics on the basis of a four-step binding model (Adair scheme), with four on-rates and four off-rates, the first and fourth of which are given in Table 5. It is instructive to note that the kinetic parameters for "R"-state Hb are very similar to those of Mb and the isolated subunits, while the much lower O_2 affinity of "T"-state Hb stems from a dramatically *increased off-rate*.

With the advent of simple porphyrin systems capable of reversibly binding O_2 in the absence of a protein, kinetic measurements on model compounds became possible. The first such work was reported by Basolo and collaborators (129, 130), who studied O_2 and CO binding to $Fe^{II}TPP(B)_2$ complexes with B = pyridine, 1-MeIm, and piperidine. Competition between excess axial base and O_2 for the sixth coordination site added a complication that is not found in the protein systems, which enforce five-coordination in the deoxy state. Mixing solutions of $FeTPP(B)(O_2)$ with large excesses of B (at $-79°C$ to prevent oxidation of the O_2 complex)

Table 5 Kinetics of O_2 Binding to Hemoproteins and Model Compounds

System	T (°C)	O_2 On-rate k' ($M^{-1} sec^{-1} \times 10^{-7}$)	O_2 Off-rate k (sec^{-1})	$P_{1/2}(O_2)^c$ (torr)	Reference
Sperm whale Mb[a]	21.5	1.9	11	0.51	120
Human Hb chains[a]	20	4.8	28	0.46	120
Human Hb chains[a]	20	6.5	16	0.40	120
Human Hb (T)[a]		0.9	1080	85	122
Human Hb (R)[a]		4.0	48	0.8	122
leghemoglobin[a]	25	15	11	0.045	128
23^b (Im, 7Am)[d]	22	6	35	0.32	131
25^b (Im, 6Es)	20	4.9	160	1.8	132
26^b (Im, 7Am)	20	2.2	23	0.57	132
27^b (Im, 8Am)	20	2.9	24	0.45	132
29^b (Py, 7Es)	20	1.7	380	12.2	132
30^b (Py, 7Es)	22	1.4	150	6.1	131

[a] 0.1 M Phosphate buffer, pH 7.
[b] H_2O, pH 7.3 Phosphate buffer, 2% cetyltrimethylammonium bromide.
[c] Calculated from k'/k taking solubility of O_2 in H_2O of 1.82×10^{-6} mole/l at 1 torr pressure.
[d] Im, 7Am represents an amide linkage 7 atoms long connecting an imidazole base to the porphyrin; Py is a pyridine base and Es, an ester linkage.

gives displacement of O_2 at a rate independent of the concentration of B. This suggests that the rate-limiting step is O_2 dissociation. Conversely, $FeTPP(B)_2$ solutions react with large excesses of O_2 at a rate independent of the O_2 concentration, suggesting a rate-limiting dissociation of B to an intermediate five-coordinate complex. The overall mechanism in Eqs. (5) and (6) was therefore proposed:

$$FeTPP(B)_2 \underset{k_{-1}}{\overset{k_1}{\rightleftharpoons}} FeTPP(B) + B \qquad (5)$$

$$FeTPP(B) + O_2 \underset{k_{-2}}{\overset{k_2}{\rightleftharpoons}} FeTPP(B)(O_2) \qquad (6)$$

The two rate constants determined from the experiments above are k_{-2} and k_1. The constant k_{-2} is seen to be equivalent to the off-rates determined for the protein systems; but as Basolo points out, no quantitative comparisons can be made because of the low temperature at which the model system was necessarily studied. By measuring the overall equilibrium constant $K_{O_2} = k_1 k_2 / k_{-1} k_{-2}$ for the model system, it was possible to determine the ratio k_2/k_{-1} utilizing the observed values of k_1 and k_{-2}. This ratio is a "discrimination factor" which represents the relative *kinetic* preference of the five-coordinate complex for B versus O_2. For 1-MeIm, $k_2/k_{-1} = 1.6$, leading to the interesting conclusion that the deoxyporphyrin has only a slight kinetic preference for O_2 over 1-MeIm. Basolo notes therefore that if the distal imidazole in the proteins *could* bind to the heme, O_2 binding ability would be seriously impaired.

Traylor and co-workers recognized that by using five-coordinate model porphyrins with covalently attached bases, the difficulties encountered by Basolo involving competition of excess base for the sixth coordination site could be avoided. Employing the same flash photolysis and rapid-mixing techniques used in studies of the proteins, they directly determined the O_2 on- and off-rates and $P_{1/2}(O_2)$ values given in Table 5 (131–133). The structures of the porphyrins used are illustrated in Figure 16. As with the natural systems, the on-rates are always found to be second order while the off-rates are first order. In aqueous micellar solvent systems, the model compounds **23**, **26**, and **27** (which have an imidazole base attached through relatively unstrained linkages) display rate constants which are very close to those of Mb, isolated α and β chains of Hb, and Hb itself in the "R" state. Shortening the chain holding the imidazole as in model compound **25** results in an increased off-rate and higher $P_{1/2}(O_2)$. The investigators (132, 133) propose that a "tension" on the iron imposed by the short linkage to the imidazole causes this increased off-rate. Al-

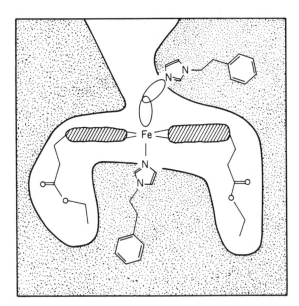

Figure 17. Wang's polymer-encapsulated heme system.

though the increase in the off-rate between the "unstrained" models and the "strained" model **25** is only a factor of about 6 while the increase between "R"- and "T"-state hemoglobin is a factor of roughly 25, the model studies are strongly suggestive of an important role for protein-induced constraints on the proximal imidazole in the mechanism of cooperativity (see further discussion in Section 4.4).

Chang and Traylor (131) have also investigated the kinetics of O_2 binding to **23** and **30** in several solvents of differing polarity (e.g., toluene, CH_2Cl_2, DMF). In general, increased polarity has little effect on the on-rates but leads to substantially decreased off-rates resulting in overall enhanced O_2 binding. This effect is ascribed to stabilization of the polar Fe—O—O linkage by more polar media. Finally the pyridine base porphyrins **29** and **30** bind oxygen more weakly than the imidazole base compounds investigated. While this is expected from studies on Fe(II) porphyrin systems with free bases in solution, quantitative comparisons of the on- and off-rates for the pyridine and imidazole base-attached porphyrins should not be made because of the steric differences resulting from dissimilar linkages.

4.3 Polymer-Supported Systems

The role of the protein in preventing irreversible oxidation of Fe(II) in oxygen transport hemoproteins is at least twofold, as we have noted in Section 3.3. First, the protein isolates individual heme units and prevents oxidation via μ-peroxo intermediate. Second, the protein protects the oxygen binding site from acids which would promote oxidation. Wang was the first to recognize and demonstrate (134, 135) that a synthetic polymer might fill these two roles of the protein. He reduced a benzene solution of Fe(III) protoporphyrin IX diethyl ester (Fe^{III}PPIXDEE) and excess 1-(2-phenylethyl)imidazole (PEIm) with aqueous dithionite under CO, then added polystyrene to the organic layer and evaporated the benzene under CO to leave a polystyrene film containing Fe^{II} PPIXDEE(CO)(PEIm). On the basis of visible spectra, Wang found, rather surprisingly, that the CO could be removed by flushing the film with N_2 for several hours at room temperature. Oxygenation in O_2 or air resulted in "an oxy-hemoglobin-type of spectrum," which could then be reversed to a deoxy spectrum under N_2. Wang's conception of the deoxy film is represented schematically in Figure 17. A similar film containing Fe^{II}PPIX showed only oxidation with O_2, probably because of the acidic side chains on the heme.

Since this initial work a number of studies have been carried out on oxygenation of heme/polymer systems. These studies can be divided into three categories: systems in which neither the porphyrin nor the axial base is covalently attached to the polymer (134–136), systems in which the axial base is covalently attached to the polymer (28, 137–146), and systems in which the heme itself or the heme and the base both are covalently attached to the polymer (147, 148).

Examples of systems of the first type include the work of Wang just discussed as well as more recent studies by Chang and Traylor (136) which utilized the heme **23** (see Fig. 16). Because this heme has a built-in imidazole, no additional axial base was required. A benzene/$CHCl_3$ solution of **23** was generated by reduction of the Fe(III) precursor with aqueous dithionite, polystyrene was added, and the organic solvent was evaporated under CO to form a polymer film of the carbonmonoxy heme. The CO was removed under an argon stream by heating to 80°C. Reversible oxygenation could then be observed spectrophotometrically, as in the earlier experiments of Wang. Chang and Traylor also obtained identical results using solid films of **23** without polystyrene.

Systems of the second type, with the base attached to the polymer, were first studied by Collman and Reed (28), who prepared a polystyrene resin containing imidazole units by treating chloromethylated polystyrene with lithium imidazolate. Treatment of this resin with $Fe^{II}TPP$ in benzene gave predominantly a six-coordinate bisimidazole heme (as judged by its visible spectrum and diamagnetism). In the presence of O_2 the heme rapidly formed $(Fe^{III}TPP)_2O$, suggesting that the solvent-swollen polymer matrix was too mobile to prevent interaction of two heme units leading to oxidation.

Tsuchida and co-workers have studied the interaction of O_2 with Fe(II) protoporphyrin IX supported on poly(4-vinylpyridine), poly(4-vinylpyridine) partially quaternized with benzyl chloride (~20%), and poly(N-vinyl-2-methylimidazole) (137–141). In aqueous buffer solutions, DMF/MeOH, and DMF/H_2O mixtures, the heme was reduced using a 100-fold excess of $Na_2S_2O_4$, which remained in the polymer solution or film through subsequent experiments. From analysis of spectrophotometric data, it is suggested that the heme–polymer complexes are predominantly five-coordinate under inert atmosphere. However using the same techniques Tsuchida also claims that simple pyridine/heme solutions (both Fe(II) and Fe(III)) remain five-coordinate with up to a 10,000-fold excess of pyridine (139), in clear contradiction to all other studies of binding of axial nitrogen bases to ferrous and ferric heme (see Sections 3.1 and 3.2). The results on the polymer solutions should accordingly be viewed with caution.

Upon admission of O_2, a product is formed which displays a broad Soret absorption around 405 nm, similar to that formed by solutions of Fe(III) heme in the corresponding polymer (144). The Soret band of oxy-Mb is found at ~418 nm and is fairly sharp. The spectral changes in the polymer solutions could not be reversed by degassing, although subsequent addition of CO led to changes ascribed to formation of a reduced CO complex. The stoichiometry of O_2 uptake was reported to be 0.5 O_2/heme, although it is not clear how this number was obtained in the presence of a 100-fold excess of $Na_2S_2O_4$, which itself reacts rapidly with O_2 in solution. The rate of reaction of the heme/polymer solutions with O_2 was found to be second order in Fe(II) and first order in O_2 (139). Although Tsuchida claims throughout this work (137–139) that dioxygen complexes are being formed, the above evidence seems to point strongly to the conclusion that oxidation is occurring.

Tsuchida has also studied the reaction of O_2 with solid poly(4-vinylpyridine)/heme (140) and with aqueous solutions of poly-l-lysine/heme (141). Visible spectra and magnetic susceptibilities are presented as evidence of reversible oxygenation in the solid polymers. With the poly-l-lysine/heme solutions, plots of the initial rate of oxygen reaction versus O_2 pressure are sigmoidal, suggesting some type of cooperative behavior. True equilibrium data could not be obtained because of the irreversible nature of the reaction, and, as with the other studies, the presence of a large excess of $Na_2S_2O_4$ complicates interpretation.

A careful and complete recent study by Fuhrhop and co-workers (144) has shed considerable light on the question of reversible oxygenation in imidazole polymer/heme systems. These investigators prepared copolymers of 1-vinylimidazole, 2-methyl-1-vinylimidazole, or 2-phenyl-1-vinylimidazole with styrene, varying the percent imidazole from 1% to 100%. Heme complexes ($Fe^{II}OEP$ and $Fe^{II}PPIXDME$) were prepared in benzene solution by reduction with aqueous dithionite, and excess dithionite was removed with the aqueous layer. Magnetic susceptibility and visible spectral results are fully consistent with the conclusion that the heme is six-coordinate in the imidazole polymer with >10% imidazole units and five-coordinate with <10% imidazole or with the 2-methyl- and 2-phenylimidazole polymers. In the solid state the low-spin hemes did not react with O_2, while the high-spin hemes clearly showed reversible oxygenation. In solution all the systems investigated underwent irreversible oxidation.

A quite different type of polymer support has recently been examined by Allcock and co-workers (146), who prepared the imidazole-functionalized polyphosphazene polymer **31** shown in Figure 18. Aqueous solutions of $Fe^{III}PPIX$-containing polymer **31** were reduced with excess dithionite. Mössbauer and electronic spectra indicated that a six-coordinate bisimidazole heme was formed. Upon exposure to O_2, rapid and irreversible oxidation of the heme occurred. In an attempt to stabilize the O_2 complex, films were prepared by evaporating the water from the ferrous

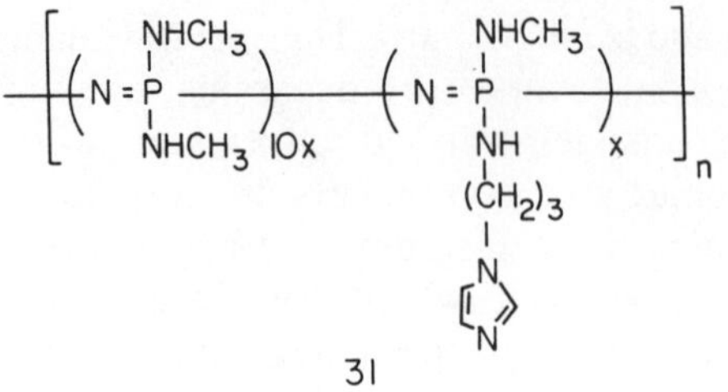

Figure 18. Attachment of imidazole ligands to polymeric supports.

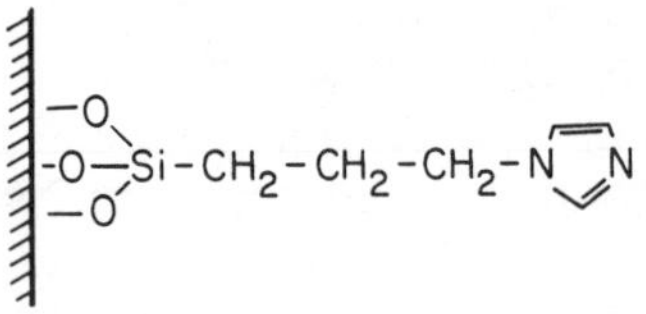

heme/polymer solutions. These films did indeed exhibit partially reversible spectral behavior on cycling between and O_2 atmosphere and vacuum. However Mössbauer spectra revealed that oxidation was occurring under O_2 and that the resulting Fe(III) species was being rereduced under vacuum by either the polymer itself or excess dithionite. *Recognition of this problem should serve as further warning that characterization of reversible dioxygen complex formation should never be made solely on the basis of visible spectral changes.*

A final example of a system in which the axial base is attached to a polymeric matrix comes from work of Basolo and co-workers (145). Recognizing that a more rigid support might be more effective at preventing oxidation via a μ-peroxo dimer, these authors prepared the imidazole-functionalized silica gel **32** shown in Figure 18. The gel typically was stirred with a solution of Fe(TPP)(pyridine)$_2$ or Fe(TPP)(piperdine)$_2$ dissolved in benzene, the solvent was decanted, and the gel washed with solvent and dried. This technique avoids contamination by excess reducing agent since the starting material is a solid, well-characterized ferrous porphyrin. The silica gel-supported porphyrins are believed to be six-coordinate initially, with the imidazole of the gel at one axial site. Heating the gel to 250°C under flowing helium removes the pyridine or piperidine, leaving a five-coordinate Fe(II) porphyrin. The resulting silica gel was found to chemisorb ~1 mole O_2 per iron at $-127°$ and to release O_2 quantitatively at room temperature. Determination of the stoichiometry was complicated by the comparitively large amount of physisorption of O_2 by the silica gel itself, which had to be factored out. Because of the low concentrations of porphyrin supported on the gel, physisorption by the gel typically accounted for two thirds of the total volume of oxygen adsorbed, forcing a rather large correction. The surprisingly weak binding of oxygen by the supported heme also made measurements difficult; $P_{1/2}(O_2)$ at 0°C was estimated to be 230 torr, compared to 0.14 torr for human Mb.

The third general class of heme/polymer systems listed at the beginning of this section is that in which the heme itself is covalently attached to the polymer. Fuhrhop was the first to adopt this approach (144). He prepared a set of terpolymers by free-radical polymerization of styrene, various 1-vinylimidazoles, and iron protoporphyrin IX dimethyl ester. The vinyl side chains of the porphyrin were thus incorporated into the backbone of the polymer. The behavior of these terpolymers was identical to the behavior of the styrene/1-vinylimidazole copolymers with complexed heme already discussed; solid terpolymers with >10% imidazole units contain six-coordinated bisimidazole heme which does not react with dioxygen, while those with <10% imidazole units contain five-coordinated heme which reacts reversibly with O_2. In solution irreversible oxidation by O_2 always occurs fairly rapidly.

A similar approach has recently been reported by Tsuchida (148). Several porphyrins with vinyl functional groups were prepared and copolymerized with styrene (e.g., 5-mono(p-acrylamidophenyl)-10,15,20-triphenylporphine). Terpolymers were also prepared by including 1-vinylimidazole in the polymerizations. The Fe(III)-containing polymers were reduced in nonaqueous solutions with Cr(acac)$_2$ and reaction with O_2 in solution monitored by visible spectrophotometry. In all cases reported the spectra are consistent with the formation of dioxygen complexes, with half-lives at 20°C on the order of 4 to 40 minutes.

A somewhat more complex approach was adopted by Bayer and Holzbach (147), who functionalized the two polymers **33** and **34** in Figure 19 with imidazole and porphyrin units, utilizing peptide synthesis techniques. A typical resulting polymer **35** is also illustrated in Figure 19. Reduction of the iron with dithionite, followed by chromatographic removal of excess reductant, leads to a polymer that shows "reversible" spectral changes in aqueous solution on cycling between an oxygen atmosphere and vacuum, though after several cycles oxidation occurs. As no further characterization of the O_2 complex was reported, the likelihood that "reversibility" is due to reduction of oxidized iron as observed by Allcock (146) must be considered. As in Tsuchida's work with poly-*l*-lysine polymers (141), sigmoidal oxygen binding curves are reported by Bayer and Holzbach.

Covalent attachment of a porphyrin to amine-containing polymers has also been studied by Ledon and Brigandat (149). These workers prepared

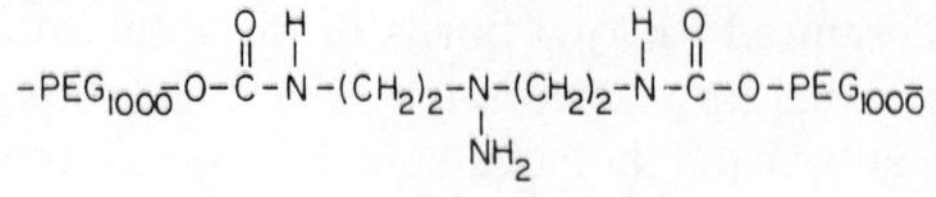

Figure 19. Bayer and Holzbach's polymer-bound heme systems.

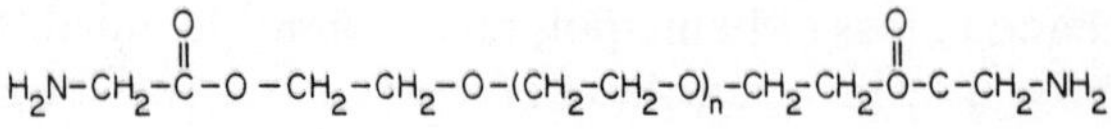

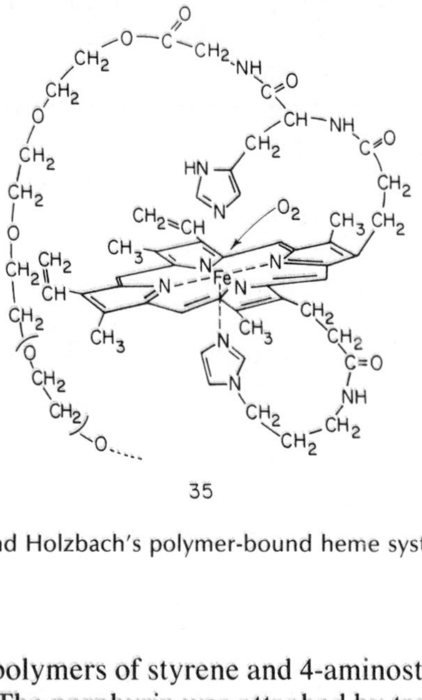

macroporous copolymers of styrene and 4-aminostyrene crosslinked with divinylbenzene. The porphyrin was attached by treatment of the polymer with tetra(p-chlorocarboxyphenyl)porphyrin, and Fe(II) was inserted directly with $FeCl_2$ in DMF. When solid polymer crosslinked with 20% divinylbenzene is exposed to air, irreversible oxidation occurs and a characteristic μ-oxo infrared band is observed at 860 cm^{-1}. However with 30% divinylbenzene-crosslinked polymer reversible oxygenation is obtained, as judged by visible spectra and the absence of the μ-oxo IR band.

In summary, there appears to be some evidence for relatively stable dioxygen complexes of Fe(II) porphyrins in solid polymers, particularly in cases where the polymer is fairly rigid and the presence of acidic groups is minimized. Convincing proof is still lacking however in the majority of cases. In solution or solvent-swollen polymers there is little or no good

DIOXYGEN BINDING TO IRON PORPHYRINS

evidence for the existence of a stable dioxygen complex of an Fe(II) porphyrin.

4.4 Protected Pocket Porphyrins

4.4.1 Synthesis. During the last several years a number of macrocycles have been synthesized in which access to the metal is restricted by the presence of bulky peripheral substituents. It was hoped that these "protected pocket" ligands would greatly decrease the tendency to form the more stable μ-peroxo and μ-oxo dimers and thus stabilize the 1:1 O_2/Fe complexes (Fig. 20). The first attempted synthesis of such a porphyrin is due to Traylor (150); unfortunately this porphyrin could be prepared in only very poor yield, and no chemical studies were possible. Baldwin made the first report (112) of reversible oxygenation of the Fe(II) complex of a "pocketed" macrocycle (see Fig. 15). In retrospect however the steric encumberance of this compound appears to be insufficient to prevent oxidation; rather, it was the reduced temperature ($-85°C$) which provided most of the kinetic stability of the oxygen adduct. A number of other protected porphyrins suffer similar difficulties. The strapped porphyrins of Ogoshi (51), Baldwin (48), and Battersby (50), which have a single hydrocarbon chain linked over the face of the porphyrin, do not support reversible oxygenation of their Fe(II) complexes, apparently because the strap can be pushed out of the way of an incipient μ-oxo dimer. Likewise, the bis-pocketed porphyrin of Vaska apparently does not prevent irreversible oxidation, though the reports are sketchy (151, 152).

To date there are only three porphyrins which provide sufficient steric protection to stabilize the O_2 complex at room temperature for any length of time. Collman et al. prepared the "picket fence" (114, 153) and "tailed picket fence" porphyrins (45, 46) shown in Figure 21. Baldwin's elegant synthesis of the "capped" porphyrin (47) utilizes a pyromellitic acid-derived tetraaryl aldehyde (Fig. 10). The ferrous complex of this porphyrin will reversibly bind O_2 if a large excess of axial base (e.g., 1-methylimidazole, pyridine) is present. Unfortunately the least soluble component of the complexes present is the four-coordinate species, and Baldwin has

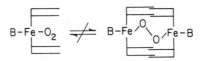

Figure 20. The "protected pocket" approach to stabilizing iron(II) dioxygen complexes.

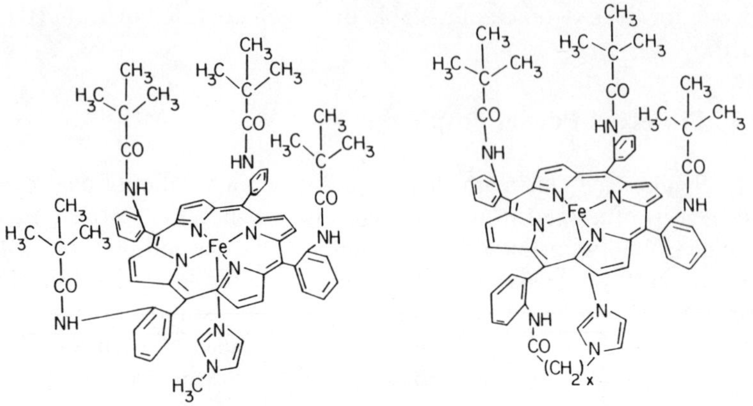

FeTpivPP (1-MeIm)

x = 3 : FePiv₃(4ClmP) Por
x = 4 : FePiv₃(5ClmP) Por

Figure 21. Picket fence and tailbase picket fence porphyrins.

been unable to isolate either five- or six-coordinate (e.g., O_2 adduct) compounds in this system.

The synthesis of the "picket fence" porphyrins is based on the biphenyl-like atropisomerism of ortho-substituted *meso*-tetraphenylporphyrins, which was first noted by Ullman (155). These porphyrins have phenyl groups oriented nearly normal to the porphyrin plane, projecting the ortho substituent above or below the plane. Even for rather small substituents (such as —OH or —NH₂), the various atropisomers are stable enough to be separated by chromatography. The synthesis of the "picket fence" porphyrin [which we have abbreviated H₂TpivPP for *meso*-tetra(α,α,α,α-*o*-pivalamidophenyl)porphyrin] has been reported in detail (114). Briefly, the condensation of *o*-nitrobenzaldehyde with pyrrole yields *meso*-tetra(*o*-nitrophenyl)porphyrin, which is then reduced with stannous chloride to a mixture of *meso*-tetra(*o*-aminophenyl)porphyrin atropisomers; the *meso*-tetra(α,α,α,α-*o*-aminophenyl)porphyrin can be isolated from this mixture by chromatography. The reaction of this atropisomer with excess pivaloyl chloride, (CH₃)₃COCl, produces H₂TpivPP, the "picket fence" porphyrin. If the tetraamino compound is mixed with three equivalents of pivaloyl chloride, chromatographed, and heated to rotate the remaining aminophenyl ring and rechromatographed, the *meso*-tri(α,α,α-*o*-pivalamindophenyl)-β-*o*-aminophenylporphyrin (a "three picket" compound) can be isolated. The remaining amino group, which has been separated in the down (i.e., β) orientation relative to the up (i.e., α) conformation of the pivalamido pickets, can be treated with acid chlorides bearing other functional groups (e.g., imidazole, thioether, thiol, or pyridine) to yield "tailed picket fence" porphyrins (45, 46), as shown in Figure 21. The ferrous complexes of these porphyrins are obtained either by reduction of the ferric species (114) or by the insertion (46) of Fe(II) from $FeBr_2$ in benzene/tetrahydrofuran/2,6-lutidine under N_2.

4.4.2 Physical Properties of Iron Porphyrinate–O_2 Complexes. The physical properties of these synthetic analogs have been extensively compared with those of the hemoproteins. Nearly every imaginable physical technique has been used to probe myoglobin and hemoglobin. Electronic, vibrational, and Mössbauer spectroscopies however have proved most useful for comparison of the proteins and the model compounds. X-Ray diffraction studies of the synthetic analogs have provided detailed structural information about both the deoxy and oxy forms. In general the physical properties of the proteins and the analog compounds are quite similar. In addition, as shall be discussed later, the O_2-binding properties are also similar.

Electronic Spectra. Electronic spectra provide a convenient means of comparing the porphyrins with oxyhemoproteins. These electronic spectra are dominated by the $\pi \rightarrow \pi^*$ transitions of the porphyrin and are perturbed only slightly by the metal, its axial ligands, and porphyrin substituents. The phenyl meso-substituents of the "capped" and "picket fence" porphyrins do change the electronic spectra from biological porphyrins generally by red shifting the bands 5 nm. Nonetheless, as Table 6 shows, the spectral similarities between the hemoproteins and the synthetic analog are recognizable.

Vibrational Spectra. A more satisfactory comparison of the iron-dioxygen species in Mb or Hb and in "picket fence" metalloporphyrins can also be developed from the vibrational spectra. Resonance Raman spectroscopy provides the Fe—O stretching frequency, while infrared absorption yields the O—O stretching frequency. The relevant data are given in Table 7. The Fe—O stretching frequency is not significantly coupled with other atomic motions, as indicated by isotopic labeling. The difference between Fe—^{16}O and Fe—^{18}O vibrational frequencies is predicted by a diatomic harmonic oscillator calculation to be 25 cm^{-1}, very close to that observed in both oxy-Hb and in (O_2)FeTpivPP(1-MeIm). Thus the excellent agreement between the protein and the synthetic analogs indicates that the Fe—O bonding is quite similar. The O—O stretch is also similar in the O_2 adducts of Hb, Mb, and the "picket fence" porphyrins.

Table 6 Visible Spectra of Hemoproteins and Analogs from 400 to 600 nm[a]

Compound	Soret Band	α Band	β Band	Reference
Five-coordinate				
Mb (sperm whale)	434 (5.1)	--	556 (4.1)	7
Hb (human)	430 (5.1)	--	555 (4.1)	7
FeTpivPP(1,2-Me$_2$Im)	437 (5.3)	538 (3.8)	559 (3.9)	45
FePiv$_3$(5ClImP)Por	438 (5.3)	536 (3.9)	558 (3.9)	45
FePiv$_3$(4ClImP)Por	438 (5.3)	537 (3.9)	559 (3.9)	45
O$_2$Adducts				
MbO$_2$	418 (5.1)	543 (4.1)	581 (4.2)	7
HbO$_2$	415 (5.1)	541 (4.1)	577 (4.2)	7
(O$_2$)FeTpivPP(1-MeIm)	426 (5.2)	548 (4.2)	582 (3.4)	114
(O$_2$)FePiv$_3$(5ClImP)Por	425 (5.1)	546 (4.1)	585 (3.4)	45
(O$_2$)FePiv$_3$(4ClImP)Por	426 (5.1)	548 (4.1)	585 (3.6)	45
(O$_2$)Fe(capped Por)	434	545	580	154
CO Adducts				
MbCO	423 (5.3)	542 (4.1)	579 (4.1)	7
HbCO	419 (5.3)	540 (4.1)	569 (4.1)	7
(CO)FePiv$_3$(4ClImP)Por	425 (5.5)	540 (4.1)	579	45
(CO)FePiv$_3$(5ClImP)Por	424 (5.5)	540 (4.1)	579	45

[a] λ_{max} in nm (log ε, molar extinction coefficient); proteins in H$_2$O, analogs in C$_6$H$_6$ or C$_7$H$_8$.

The change in the observed frequency predicted by a diatomic harmonic oscillator calculation for $^{16}O_2$ to $^{16}O-^{18}O$ is 33 cm^{-1} and for $^{16}O_2$ to $^{18}O_2$ is 67 cm^{-1}. That these predicted frequencies are found in some cases but not in others is indicative of a weak Fermi coupling with other vibrations (157). This coupling causes shifts of some 30 cm^{-1} from the predicted frequencies for the proteins and some of the analogs and may explain some of the difference between the O—O stretching frequencies of the proteins and the analogs shown in Table 7; other contributions may include differences in local polarity or steric interactions with the O$_2$. Regardless, the observed frequencies are all in the narrow range observed for the "superoxo" type ligand (20, 164). It is clear from the data in Table 7 that this O—O frequency is quite insensitive to choice of macrocycle, axial ligand, metal, or O$_2$ affinity. The same should not be true of the Fe—O stretch; work is in progress with T. G. Spiro.

Mössbauer Spectra and Magnetic Susceptibilities. The Mössbauer spectra of Mb and Hb have been extensively investigated (165) and provide another useful comparison with the "picket fence" porphyrin complexes. Whereas electronic spectra reflect perturbations in the porphyrin π system and vibrational spectra those of the Fe–O_2 potential surface, Mössbauer spectroscopy probes the electron density at the iron nucleus. Since this s electron density is indirectly affected by shielding from p and d electrons, Mössbauer spectroscopy has proved a sensitive measure of changes in the overall distribution of electron density in the vicinity of the iron nucleus. In the absence of a magnetic field, two parameters are needed to characterize a spectrum: the chemical isomer shift (δ), which reflects the amount of electron density at the nucleus, and the quadrupole splitting ΔE_Q, which reflects the extent of spatial asymmetry in that density. Table 8 summarizes the reported data. An interesting aspect of the hemoprotein Mössbauer spectra is the unusual temperature dependence of ΔE_Q in HbO_2, an $S = 0$ system; the temperature dependence is even greater in (O_2)FeTpivPP(1-MeIm). This has been attributed to a dynamic equilibrium of the bent Fe—O—O unit as the uncoordinated O atom swivels relative to the plane of the coordinated imidazole (169). The usual explanation for such temperature dependence of ΔE_Q involves changes in the relative populations of thermally accessible paramagnetic states. This explanation was not considered acceptable for HbO_2 which was presumed to be diamagnetic at all temperatures (169). Recently however Cerdonio and co-workers (171, 172) have studied the magnetic susceptibility of HbO_2 from 25° to 285°K and have found, contrary to Pauling and Coryell's classic experiments (173), that HbO_2 does indeed have a thermally populated paramagnetic excited state some 150 cm^{-1} above the

Table 7 Infrared and Resonance Raman Spectra of Superoxo Complexes

Complex	$\nu_{Fe-O}{}^a$	$\nu_{O-O}{}^a$	Reference
MbO_2	567 (540)	1107 (1065)	156
MbO_2	—	1103 (1065)	157
(O_2)FeTpivPP(1-MeIm)	568 (545)	1159 (1075)	63, 158
(O_2)FeTpivPP(1-TrIm)	—	1163 (1080, 1127)	63
(O_2)FeTpivPP(2-MeIm)	—	1158 (1093, 1121)	159
(O_2)FeTpivPP(1,2-Me$_2$Im)	—	1159 (1093, 1122)	159
$CoHbO_2$	—	1105 (1065)	160
(O_2)CoTpivPP(1-MeIm)	—	1150 (1077)	63
(O_2)CoTpivPP(1-TrIm)	—	1153 (1077, 1125)	63
(O_2)Co(bzacen)(py)	—	1128	161
$[(O_2)Co(CN)_5]^{3-}$	—	1138	162
(O_2)CrTPP(py)	—	1142	163

^{a} Frequencies in cm^{-1} for $^{16}O_2$ ($^{18}O_2$, ^{16}O–^{18}O) adducts.

Table 8 Mössbauer Spectral Data of Hemoproteins and Analogs

Compound	Isomer Shift δ (mm/sec ± 0.01)			Quadrupole Splitting ΔE_Q (mm/sec)			Reference
	4.2°K	77°K	195°K	4.2°K	77°K	195°K	
Mb	0.92	0.91	0.86	2.22	2.17	1.78	166
Hb	0.93	0.92	0.89	2.26	2.22	1.94	167, 168, 170
FeTPP(2-MeIm)	0.93	0.91	0.87	2.28	2.26	1.97	114
FeTpivPP(1-MeIm)		0.88	0.84		2.32	2.01	114, 20
FeTPP(1,2-Me$_2$Im)	0.92	0.91	0.86	2.16	2.04	1.70	170
HbO$_2$	0.24	0.26	0.20	2.24	2.19	1.89	169
(O$_2$)FeTpivPP(1-MeIm)	0.28	0.27	0.24	2.11	2.04	1.39	114, 169
(O$_2$)FeTpivPP(1-nBuIm)	0.28	0.26	0.24	2.10	1.70	1.31	114, 169

diamagnetic ground state. This result has been disputed (174) and reaffirmed (172). The Mössbauer spectrum of (O$_2$)FeTpivPP(1-MeIm) at 4.2°K in a strong transverse magnetic field confirms that the ground state is diamagnetic but does not give any information about thermally accessible excited states (169). Thus the Mössbauer spectroscopy of hemoglobin and of the "picket fence" O$_2$ adducts is consistent with (but does not require) a low-lying paramagnetic excited state.

The magnetic susceptibility measurements of (O$_2$)FeTpivPP(1-MeIm) only partially clarify the situation. At room temperature, freshly prepared samples have an observed moment of 0.5 to 1.0 Bohr magneton (BM), corresponding to a small amount of Fe(III) impurity (respectively 0.7 to 3% high-spin ferric) (20). In crystalline samples of (O$_2$)FeTpivPP(1-MeIm) the magnetic susceptibility over the range of 5° to 300°K shows nearly ideal Curie behavior, as expected for <3% high-spin Fe(III) impurity, the presence of which is confirmed qualitatively by the EPR spectra (45). The magnetic susceptibilities have been measured by Evan's method (175) in solution. Using a 100 MHz NMR spectrometer, no discernible shifts were observed (45) in the standard for either (O$_2$)FePiv$_3$(5ClmP)Por or (CO)FePiv$_3$(5ClmP)Por, which implies that μ < 2 BM. Cerdonio's results on HbO$_2$ however indicate (172) a magnetic moment of 2.4 ± 0.3 BM at 285°K, which is much greater than that observed for (O$_2$)FeTpivPP(1-MeIm). If the excited triplet state of the analogs were just another 100 cm^{-1} or so above that in Hb, the magnetic susceptibility data available would not reveal it. The application of infrared magnetic resonance spectroscopy (176, 177) to HbO$_2$ and synthetic analogs might produce direct evidence of the existence of a thermally accessible paramagnetic state.

DIOXYGEN BINDING TO IRON PORPHYRINS

This technique detects the field-dependent transitions in the energy range of roughly 5 to 500 cm^{-1} and has been used previously to study electron-spin state transitions in the presence of large zero field splitting.

Structural Data. Crystal structures have been determined for a large number of various derivatives of hemoproteins and model compounds. In this section we discuss only those of the oxygen adducts and compare them to the deoxy complexes. The structural parameters have been summarized in Table 9. With the exception of deoxyerythrocruorin, there is good agreement between the proteins and the synthetic five-coordinate Fe(II) porphyrinates that the Fe is drawn out of the mean plane of the porphyrin toward the coordinated imidazole by ~0.5 Å. As discussed in Section 3.2, the porphyrin "domes" somewhat to accommodate this geometry, but the extent of this doming varies depending on the constraints on the porphyrin. The deoxyerythrocruorin structure shows an anomalously small out-of-plane displacement of the iron (184); this structure is only in a preliminary stage of refinement however.

For many years the crystal structures of oxyhemoproteins eluded solution because of oxidation during collection of diffraction data (185). The first structure of an O_2 complex of an iron porphyrinate was in fact that of the "picket fence" porphyrin, (O_2)FeTpivPP(1-MeIm). In this complex, the iron atom lies essentially in the mean plane of the porphyrin (in fact, displaced slightly, 0.03 Å, *toward,* the O_2); the O_2 is bound in an end-on fashion with an Fe—O—O angle of <131°. Because of disorder and large thermal motion of the terminal oxygen atom, only a lower bound for the O—O distance and an upper bound for the Fe—O—O angle could be determined (182).

Using the sterically hindered 2-methylimidazole ligand, Collman and co-workers isolated single crystals of a five-coordinate, deoxy-Fe(II) "picket fence" porphyrin complex which could be oxygenated without loss of crystallinity. Thus precise X-ray diffraction structures were obtained for both FeTpivPP(2-MeIm) and (O_2)FeTpivPP(2-MeIm) (Table 9), allowing a detailed examination of the structural changes which occur upon oxygenation (34). The 2-methyl group sterically restrains the imidazole and strongly influences the structure of the O_2 adduct. Most striking is the elongation of the Fe–O bond in (O_2)FeTpivPP(2-MeIm), while the Fe–N_{Im} distance is much less affected.

The motion of the Fe-bound imidazole upon oxygenation has been a matter of wide discussion for many years. Using the three structures available of "picket fence" porphyrins, FeTpivPP(2-MeIm), (O_2)FeTpivPP(1-MeIm), and (O_2)FeTpivPP(2-MeIm), we may estimate lower and upper

Table 9 Structures of Deoxy and Oxy Iron Porphyrinates[a]

Complex	Fe–Ct (Å)	Fe–P$_c$ (Å)	Fe–N$_P$ (Å)	Fe–N$_{Im}$ (Å)	Fe–O (Å)	O–O (Å)	Fe–O–O Angle	Reference
FeTPP(2-MeIm)·EtOH	0.42	0.55	2.086(4)	2.161(s)	—	—	—	20, 178
FeTpivPP(2-MeIm)·EtOH	0.399	0.43	2.072(5)	2.095(6)	—	—	—	34
Mb	0.42	0.55	2.06	2.1	—	—	—	179
Hb (av. of α and β)	—	0.6(1)	2.1(1)	2.1(3)	—	—	—	180
Erythrocruorin	—	0.17	2.02	2.2	—	—	—	181, 184
(O$_2$)FeTpivPP(1-MeIm)	−0.02	−0.03	1.98(1)	2.07(2)	1.75(2)	>1.16	<131°	182
MbO$_2$	—	0.33	—	2.1	1.9	1.4	121°	183
Oxyerythrocruorin	—	0.3	2.04	2.1	1.8	1.25	170°	184
(O$_2$)FeTpivPP(2-MeIm)·EtOH	0.086	0.119	1.996(4)	2.107(4)	1.898(7)	>1.22(2)	<129(2)°	34

[a] All values with estimated error of the last digit in parenthesis, when available. N$_{Im}$ = Coordinated N of imidazole; N$_p$ = porphyrin(pyrrole) N; Ct = center of the four N$_p$ plane; P$_c$ = center of mean porphyrin plane.

bounds of axial base motion upon oxygenation. Upon oxygenation of FeTpivPP(2-MeIm), the 2-MeIm group moves 0.30 Å toward the mean porphyrin plane, its steric bulk still restraining the iron atom to be 0.12 Å out of this plane. Thus the motion of an unhindered axial base (e.g., idine or 1-MeIm) must be at least 0.30 Å. An estimate of the upper bound to motion of an unrestrained imidazole comes from the relative position of the bases in the structures of FeTpivPP(2-MeIm) and (O_2)FeTpivPP(1-MeIm), from which the change in the imidazole–porphyrin distance is 0.48 Å. This should be an upper bound because the 2-MeIm ligand tends to be a little further from the porphyrin than the unrestrained 1-MeIm, even in deoxy structures. For example, in CoTPP(1,2-Me$_2$Im) compared to CoTPP(1-MeIm), the imidazole is 0.07 Å further from the mean plane of the porphyrin (97); since iron shows greater out-of-plane displacement than cobalt in five-coordinate complexes, this effect will be even smaller than 0.07 Å for FeTpivPP(2-MeIm). Thus, within the context of solid-state structures of "picket fence" porphyrins, *the motion of an unrestrained imidazole ligand upon oxygenation is between 0.48 and 0.30 Å toward the porphyrin plane.* One must be very cautious in applying these estimates to the hemoproteins however, since they do not take into account the role of the globin in determining porphyrin doming and imidazole orientation.

Recently two preliminary structures of oxyproteins have been reported. By crystallizing deoxy-Mb, exposing these crystals to O_2, and collecting X-ray diffraction data at $-12°C$, Phillips (183) was able to solve the structure of MbO_2. In the preliminary solution (refinement has not yet converged), the Fe—O—O unit is bent with an angle of 121°, similar to that of the "picket fence" complexes. In contrast however the iron atom of MbO_2 was found 0.33 Å *out* of the mean porphyrin plane toward the coordinated imidazole of the proximial histidine. If this preliminary analysis proves correct, the difference between Mb and (O_2)FeTpivPP(1-MeIm) may be due to the eclipsed orientation of the imidazole plane relative to the pyrrole nitrogens which is present in MbO_2, but not (O_2)FeTpivPP(1-MeIm) (183). Contemporaneously Weber et al. (184) reported the structure of oxyerythrocruorin. In this case the oxygenated crystals were prepared by gaseous reduction of the ferric form with H_2S and subsequent exposure to O_2. The resolution is extremely high (1.4 Å) for a protein structure, but again refinement has not been completed. The preliminary structure of this O_2 complex is unusual. First, there appears to be a water molecule hydrogen bonding to the bound O_2; second, the Fe—O—O unit is nearly linear (170°); and finally, the iron atom is 0.3Å out of the porphyrin plane, toward the coordinated imidazole. There are no other examples of linear M—O—O complexes in the inorganic literature (188), and several molecular orbital calculations indicate that the linear geometry is energetically unlikely (186, 187). It is conceivable that the H_2O molecule hydrogen bonded to the O_2 (which is also unique) could stabilize this configuration. More disturbing is the out-of-plane displacement of the iron atom, which is larger in the oxy than in the deoxy form! This is very difficult to understand; perhaps further refinement of the diffraction data will clarify this unusual structure.

The novel technique of extended X-ray absorption fine structure spectroscopy (EXAFS) has been applied to Hb and the "picket fence" compounds. The EXAFS technique can give bond length information about the atoms coordinated to transition metals even in solution or amorphous solids. Eisenberger et al. (38) have reported the average iron to porphyrin nitrogen distances (Fe–N_p) for Hb (2.055 ± 0.01 Å), FeTpivPP(1-MeIm) (2.055 ± 0.01 Å), HbO_2 (1.986 ± 0.01 Å), and (O_2)FeTpivPP(1-MeIm) (1.979 ± 0.01). Direct comparisons show that the Fe–N_p distances in the pairs Hb and FeTpivPP(1-MeIm) and HbO_2 and (O_2)FeTpivPP(1-MeIm) are identical to within 0.007 Å. These data agree well with the X-ray diffraction structures of FeTpivPP(2-MeIm) (Fe–N_p = 2.07 ± 0.005 Å) and of (O_2)FeTpivPP(1-MeIm) (Fe–N_p = 1.979 ± 0.01 Å). The Fe–N_p distance is not very sensitive to the position of the iron relative to the plane of the porphyrin nitrogens and even less to the mean plane of the porphyrin, and so it is difficult from the EXAFS data to estimate accurately the motion of the imidazole relative to the porphyrin upon oxygenation. It is clear however that the early estimate of the out-of-plane displacement of the iron in the deoxy form (>0.7 Å) was too large (22, 23).

4.4.3 Thermodynamics of O_2 Binding.

Comparisons of spectroscopic and structural properties are only a part of the synthetic analog studies of Mb and Hb. Though many of the properties of a biochemical system may prove intriguing, ultimately the fundamental interest must be in its functional properties. The prime questions of the next sections ask how the protein and heme interact to produce the observed oxygen storage and transport.

Stability to Irreversible Oxidation. The Fe—O_2 group is thermodynamically unstable. Even Hb in vivo has roughly a 3% concentration of the oxidized, ferric form, in spite of continued enzymatic rereduction (188). As was discussed earlier, the decomposition of iron–oxygen complexes is retarded by steric protection, low temperatures, low porphyrin concentrations, and high oxygen pressures. Protons greatly accelerate the irreversible oxidation (7, 12), presumably through transient formation of an Fe—O—O—H species with subsequent decomposition to Fe(III) and H_2O. This acid catalysis is present in both the synthetic complexes and the proteins. A number of mutant Hbs, for example, have been isolated which contain acidic residues in the binding pocket of either the α or the β chains, but *not* both [e.g., tyrosine substituted for the distal histidine as in HbM Boston's α chains or HbM Saskatoon's β chains (7), in which the mutant subunits contain nonfunctional ferric hemes.] Other lewis acids, for example Cu^{2+}, may also catalyze the oxidation (7).

The protected porphyrin complexes show the same order of stability to oxidation as Mb or Hb. In nonaqueous solution (10^{-4} M), under 1 atm O_2, the "picket fence" complex (O_2)Fe(TpivPP)(N-MeIm) is kinetically stable for prolonged periods (half-life about two months at 25°) (20), provided that ~3 equivalents of axial base are present to protect the unshielded side of the porphyrin. The "tailed picket fence" complexes, which need no free concentration of axial base, are less stable (45) (with half-lives of ~33 h at 25°). Baldwin's "capped" porphyrins are even less stable (154): the dioxygen complex has a half-life of 5 h in the presence of ~0.5 M 1-MeIm at 25°. A reasonable explanation (154) of the increased stability of the "picket fence" versus the "capped" metalloporphyrins lies in the ability of FeTpivPP to bind *two* axial ligands, forming for example FeTpivPP(1-MeIm)$_2$ in solution. The "capped" porphyrin excludes a second base from coordinating, thus raising the relative concentration of the very oxygen-sensitive four-coordinate complex and increasing the rate of oxidation. Iron(II) porphyrinate complexes with covalently linked imidazole "tails" but without sterically protected pockets completely oxidize within minutes upon exposure to O_2 (41).

The "picket fence" porphyrin complexes are porous in the solid state and show reversible oxygenation. Such solids can be prepared directly as the O_2 complexes; under vacuum these form the five-coordinate deoxy complexes (114, 189). Conversely, they can be prepared as five-coordinate complexes (in the case of sterically hindered axial bases) under inert atmosphere and then oxygenated by the addition of O_2 (34, 189). The rate of decomposition of these solids is very slow. Cycling between O_2 and vacuum causes less than 0.5% loss of activity for >100 cycles at 25° over a period of months. Nonetheless oxidation does occur slowly, as shown by a gradual increase in the magnetic susceptibility and growth of an EPR signal assignable to high-spin Fe(III).

O_2 Binding to "R"-Type Hemoproteins and Analogs. The spectroscopic and structural properties of the "picket fence" porphyrin O_2 complexes and the hemoproteins are all quite similar. More importantly, so are the thermodynamics of oxygenation. As discussed earlier, the range of O_2

affinities varies tremendously from protein to protein, indicating the dramatic influence the globin can have on the O_2 binding properties of the heme. Those hemoproteins which have O_2 affinities similar to that of "R"-state Hb (the higher-affinity, oxy form) include Mb, isolated subunits of Hb, mutant Hbs which lack cooperativity, and various nonmammalian hemoproteins. These "R"-type hemoproteins have $P_{1/2}$ values of roughly 0.5 torr at 25°, which is the midrange of all the affinities of the various hemoproteins.

We have measured the $P_{1/2}$ of oxygenation, as well as the enthalpy and entropy, for a number of "picket fence" porphyrin complexes by spectrophotometric and manometric techniques, as presented in Table 10. Comparisons are made between the Fe(II) synthetic analogs and "R"-type proteins in Table 11, and between Co(II) synthetic analogs and cobalt-substituted proteins in Table 12; the results are summarized schematically in Figure 22. Clearly the oxygen binding parameters of the iron porphyrins with unhindered axial bases (e.g., 1-MeIm or our "tailed" imidazoles) are all nearly identical: in general, $P_{1/2}$ at 25° is ~0.6 torr, $\Delta H°$ = -16.2 ± 0.6 kcal/mole (-67.7 ± 2.5 kJ/mole), and $\Delta S°$ = -40 ± 2 eu (standard state 1 atm) (7). As noted before (190), the intrinsic O_2 affinity of these simple porphyrins is the same as that of the native hemoproteins.

Table 10 Thermodynamic Values for O_2 Binding to "R"- and "T"-State Models

System	$P_{1/2}(25°C)$ (torr)[a]	$\Delta H°$ (kcal/mole)[b]	$\Delta S°$ (eu)[b,c]	Reference
FeTPivPP(NMeIm)				
Solid state	0.49	-15.6 ± 0.2	-38 ± 1	190
FePiv$_3$(4ClImP)Por				
Toluene solution	0.60	-16.8 ± 0.5	-42 ± 2	46
FePiv$_3$(5ClImP)Por				
Toluene solution	0.58	-16.3 ± 0.8	-40 ± 3	46
FeTPivPP(1,2Me$_2$Im)				
Toluene solution	38	-14.3 ± 0.5	-42 ± 2	46
CoTPivPP(1-MeIm)				
Solid state	61	-13.3 ± 0.9	-40 ± 3	90
Toluene solution	140	-12.2 ± 0.3	-38 ± 1	90
CoTPivPP(1,2-Me$_2$Im)				
Toluene solution	900	-11.8 ± 0.4	-40 ± 2	90

[a] Interpolated from $\Delta H°$ and $\Delta S°$; estimated errors ±5%.
[b] Error limits given are the standard deviations of van't Hoff plots.
[c] Standard state, 1 atm O_2.

Table 11 O₂ Affinities of Iron Porphyrins and Hemoproteins

System	Physical State	$P_{1/2}(25°)$ (torr)	Reference
Mb (sperm whale)	pH 8.5	0.70	191
Hb (human)			
α chains	pH 7.5,	0.63	
β chains	0.1 M phosphate	0.25	
Hb (human, "R")[a]	Various[b]	0.15–1.5[c]	88
FePiv₃(4CImP)Por	Toluene solution	0.60	46
FePiv₃(5CImP)Por	Toluene solution	0.58	46
FeTPivPP(1-MeIm)	Solid State	0.49	190
Hb (human, "T")[a]	Various[b]	9–160[c]	88
FeTPivPP(1,2Me₂Im)	Toluene solution	38	46

[a] These are actually the fourth and first intrinsic $P_{1/2}$ values, respectively.
[b] Imai's conditions include various combinations of 0.1 M NaCl, 0.1 M phosphate, 2 mM inositol hexaphosphate, and 2 mM 2,3-diphosphoglycerate, all at pH 7.4.
[c] The ratio of these "R" and "T" affinities also vary as a function of conditions, from around 40 to 500.

These model systems do not have any other particularly unusual characteristics (34, 182); there is no possibility of hydrogen bonding or electron pair donation to the bound oxygen, nor is the binding pocket extremely polar, nor is it shaped to fit the bound O_2. To the extent that these statements are true, we may then view the oxygen affinities of both model systems and relaxed hemoproteins (e.g., Mb, isolated-chain Hb, and R-state Hb) as originating solely from the ferrous porphyrin imidazole system. In contrast, the carbon monoxide affinities are much lower in most hemoproteins than in the model, due presumably to the steric constraints of the hemoprotein binding pocket and the biological necessity to decrease the CO affinity relative to O_2 (63).

The cobalt porphyrins (Table 10) provide strong confirmation of the simple nature of O_2 binding to "R"-type hemoproteins. The models with an unhindered imidazole (i.e., 1-MeIm) have $P_{1/2}$ values virtually identical to those of the reconstituted cobalt hemoproteins (Table 12). In addition, $\Delta H°$ and $\Delta S°$ of the analogs also compare well with that of CoMb: -12.8 ± 0.6 kcal/mole and -39 ± 1 eu for the models, compared to a range

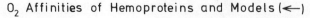

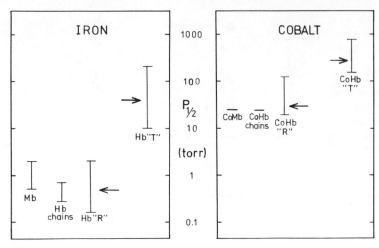

Figure 22. O₂ Affinities of hemoproteins and models.

of -13.3 to -11.3 kcal/mole for $\Delta H°$ and -40 to -33 eu for $\Delta S°$ (standard state 1 atm) (86, 194). In line with these $\Delta S°$ values is that predicted (190) on statistical mechanical grounds due to the loss of translational and rotational entropy of bound O_2. A note should be made in comparing these studies with those conducted previously on cobalt porphyrins. Earlier

Table 12   O₂ Affinities of Cobalt Porphyrins and Cobalt Hemoproteins

System	Physical State	$P_{1/2}(15°)$ (torr)	Reference
CoMb (sperm whale)	0.1 M Phosphate, pH 7	30	86, 194
CoHb (isolated chains)	0.1 M Phosphate, pH 7.4	25	195
CoHb (human, "R")[a]	Various[b]	16–180[c]	88
CoTPivPP(1-MeIm)	Solid state	28	46
	Toluene solution	70	46
CoHb (human, "T")[a]	Various[b]	120–500[c]	88
CoTPivPP(1,2-Me₂Im)	Toluene solution	450	46

[a] These are actually the first and fourth intrinsic $P_{1/2}$ values.
[b] Imai's conditions include various combinations of 0.1 M NaCl, 0.1 M phosphate, 2 mM inositol hexaphosphate, and 2 mM 2,3-diphosphoglycerate, all at pH 7.4.
[c] The ratio of these "R" and "T" affinities also vary as a function of conditions from around 2 to 16.

measurements of $P_{1/2}$ for simple cobalt porphyrins such as cobalt protoporphyrin IX dimethyl ester(1-MeIm) (105), cobalt tetra(p-methoxyphenyl)porphyrin(1-MeIm) (94, 95), and cobalt tetratolylporphyrins with appended bases (108) showed very low affinities, 300 times lower than our models or the hemoproteins. There are two explanations for this difference (90). There may be a selective solvation of the simple flat porphyrins favoring their deoxy form; or alternatively, the protein and the "picket fence" complex may provide a special stability to the O_2 adduct, through local polarity for example.

By using a series of iron and cobalt "picket fence" porphyrins, it has been possible to reproduce the O_2 affinities of Mb, Hb, CoMb, and CoHb. The basic O_2 affinity shown by model ferrous and cobaltous porphyrins is the same as that of unrestrained hemoproteins (e.g., Mb, isolated-chain Hb, R-state Hb); special interactions between the protein and the bound oxygen are not needed to explain the oxygen affinities of these hemoproteins.

In addition to the "R"-type hemoproteins, there exists an interesting class of high O_2-affinity nonmammalian hemoglobins. Included here are leghemoglobin (found in the nitrogenase system of legume nodules) and the hemoglobins of yeast and intestinal parasite worms, such as *Ascaris* and *Gastrophilus*. Table 13 compares the $P_{1/2}$ of oxygenation for a number of high-affinity hemoproteins. Two ways in which the protein could increase the heme's affinity for O_2 are to *stabilize* the oxygenated state (relative to "R"-type proteins) or *destabilize* the deoxygenated state. Stabilization of the oxygen adduct could involve hydrogen bonding to the bound oxygen, high local polarity in the binding pocket, electronic effects at the porphyrin periphery, ligation by an axial base other than imidazole, or interaction of a Lewis acid with the bound oxygen. None of these effects alone seems likely to increase the O_2 affinity by the 1.5 to 3.5 kcal/mole which is observed. On the other hand, the protein could destabilize the deoxy form by *compressing* the axial ligand (159) (thought to be a

Table 13 O_2 Affinities of High-Affinity Hemoproteins

System	Quaternary Organization	$P_{1/2}(20°)$ (torr)	Reference
Mammalian Mb	Monomer	0.5	7
Leghemoglobin	Monomer	0.04	13
Yeast (*Candida*)	Monomer	0.01	196
Ascaris	Octamer	0.0015	15, 197
Gastrophilus	Dimer	0.055	14

histidine on the basis of electronic spectra) toward the heme: the Perutz mechanism in reverse. Thus far the origin of the high O_2 affinities of these nonmammalian hemoproteins remains a matter of speculation.

O_2 Binding to "T"-State Hb and Analogs. The mechanism by which the O_2 affinity is altered and cooperativity effected in Hb is a matter of continuing controversy (1, 38). Within the confines of the Hoard–Perutz stereochemical mechanism for cooperativity (1, 198), Hb is viewed as having two alternative quaternary structures: the ligated, R state, whose O_2 affinity is essentially that of the isolated subunits, and the deoxy, T state, whose O_2 affinity is diminished. The lessened affinity of the T state is presumed to be caused by *constraint* of the proximal histidine to its unligated equilibrium position in which the iron is 0.5 Å out of the mean porphyrin plane. Upon ligation of O_2 (or other small molecules) to the T-state form, a tension develops as the iron atom moves into the porphyrin plane, attempting to drag the constrained proximal histidine with it. This T quaternary state is stabilized by direct salt bridges and hydrophobic contacts among the subunits as well as by indirect bonds between them mediated by solvent ions. Upon coordination of dioxygen molecules (roughly two per tetramer), the quaternary structure changes to the R state and the constraint of the imidazole is released. The R quaternary state is stabilized by the increased strength of the Fe—O_2 bonding, which more than overcomes the destabilization caused by the changes in intersubunit contacts. There are still questions as to whether the iron–imidazole bond is tensed in the *deoxy* T state (38) [recent resonance Raman results suggest it is not, since the Fe–imidazole stretching frequency is unaltered between R- and T-state deoxy-Hb (37b)] and as to the exact nature of the salt bridges and hydrophobic contacts (199). In addition the changes which occur in tertiary structure near the heme upon oxygenation include more than just motion by the coordinated imidazole (58, 200).

In order to probe the effect of steric restraint on O_2 binding to metalloporphyrins, sterically hindered axial bases were employed, as shown in Figure 23. In the sense that the 2-methyl group of 1,2-Me_2Im provides restraint to the motion of this axial base toward the porphyrin upon oxygenation, FeTPivPP(1,2-Me_2Im) and CoTPivPP(1,2-Me_2Im) are models for the T form of Hb and CoHb. The presence of such steric restraint with 2-MeIm and 1,2-Me_2Im is obvious from the existence of five-coordinate ferrous porphyrins and has been confirmed by the crystal structures of FeTPP(2-MeIm) (20), FeTpivPP(2-MeIm) (34), and CoTPP(1,2-Me_2Im) (197), which show that any motion of the hindered imidazole toward the porphyrin would greatly increase steric contact between the 2-methyl groups and the porphyrin. Recently this same approach has been used

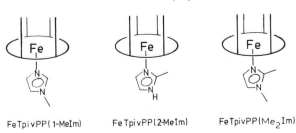

Figure 23. Picket fence porphyrins with sterically hindered axial bases.

by Traylor and co-workers in studying the kinetics of CO and O_2 binding to simple porphyrins (132).

It is interesting that the hindrance offered by the 2-methyl group of 1,2-Me_2Im is just that required to reduce O_2 affinities to the level found in T-state Hb and CoHb (Tables 11 and 12). In the iron systems the $\Delta\Delta G^0$ (the difference in O_2 affinities of metalloporphyrins with 1-MeIm and 1,2-Me_2Im) of the FeTpivPP systems is, at 25°C, 2.5 kcal/mole, which is to be compared to ΔG_{41}^0 of Hb (the free-energy difference between the first and fourth intrinsic O_2 affinities); ΔG_{41}^0 for Hb at 25° ranges from 2.1 to 3.7 kcal/mole, depending on conditions (88). In the cobalt systems, $\Delta\Delta G^0$ of the model compounds is, at 15°, 0.8 kcal/mole, and ΔG_{41}^0 for CoHb, at 15°, is 0.4 to 1.6 kcal/mole, depending on conditions (88). This correspondence is undoubtedly dependent on the degree of steric hindrance present in the model complexes, and more highly restraining bases such as 2-isopropylimidazole or 1,2,4,5-tetramethylimidazole should show even lower O_2 affinities.

The stepwise O_2 affinities of Hb shown in Table 11 have been reported by a large number of researchers. The data presented here, those of Imai et al. (88), were chosen because of the range of conditions and the existence of comparable CoHb data. Generally the data from the various groups agree well.

From Table 10 it can be noted that the restraint induced by Me_2Im in the iron porphyrins is reflected in the enthalpy of O_2 binding, the entropy remaining essentially unchanged. For the cobalt analogs the differences are much smaller but still appear to be enthalpic. This is as one would expect, since the ΔS^0 of binding is determined (190) almost exclusively by the loss of translational and rotational entropy of the O_2.

Since the metal atom is further out of the porphyrin plane for five-coordinate Fe(II) than Co(II) (69, 97, 132), one expects a greater change in the steric interaction of a bound 1,2-Me$_2$Im for the Fe(II) than the Co(II) upon oxygenation, as illustrated below.

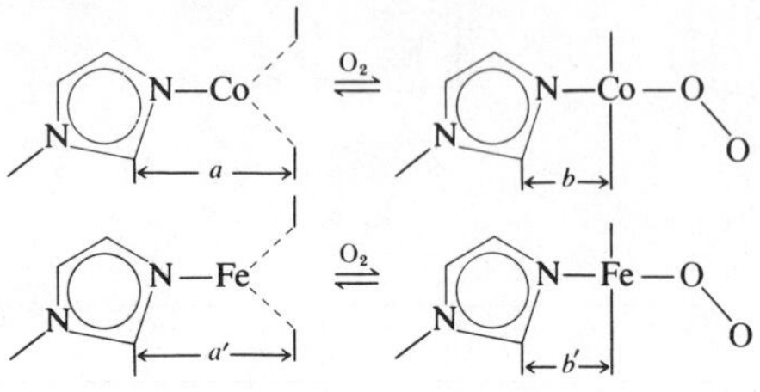

$$a < a' \qquad b \approx b'$$

That is to say, more steric interaction between the hindered axial base and the porphyrin has *already* developed in the deoxy form of Co-TpivPP(1,2-Me$_2$Im) than FeTpivPP(1,2-Me$_2$Im) decreasing the stability of MTpivPP(1,2-Me$_2$Im) relative to MTpivPP(1-MeIm) more for M = Co than Fe. Hence we expect a greater decrease in O$_2$ affinity for FeTPivPP(B) than for CoTPivPP(B) in changing B from 1-MeIm to 1,2-Me$_2$Im. Exactly this phenomenon is observed, as illustrated in Figure 22. This is completely analogous to the lessened cooperativity shown by CoHb relative to native Hb.

This same reasoning argues that even if T-state deoxy-Hb is restrained to its normal deoxy geometry (and hence unstrained) (1), then T-state deoxy-CoHb *must* be strained relative to R-state deoxy-CoHb or deoxy-CoMb, though this strain will not necessarily reside exclusively in the Co–N$_{Im}$ bond.

By tailoring the steric interactions between the axial imidazole and the porphyrin, we can mimic the decrease in ligand affinity shown in T-state Hb. The restraint presumed present in the T form of Hb and CoHb has been well modeled by FeTPivPP(1,2-Me$_2$Im) and CoTPivPP(1,2-Me$_2$Im) and provides evidence *on a molecular level* that the Hoard–Perutz mechanism is plausible.

Cooperativity in O$_2$ Binding to Iron Prophyrinates. After studying the oxygenation of hindered imidazole "picket fence" porphyrins in solution, we measured their O$_2$ binding in the solid state over a range of five orders of magnitude in O$_2$ pressure (189). These data are shown in Figure 24. The startling result is the cooperativity observed in the oxygenation of FeTpivPP(2-MeIm) and FeTpivPP(1,2-Me$_2$Im). For these two complexes there exist two regimes of noncooperative binding (at high and low O$_2$ pressures) and an intermediate region of cooperative binding. From the binding at high and low O$_2$ pressures one can extrapolate the $P_{1/2}$ values

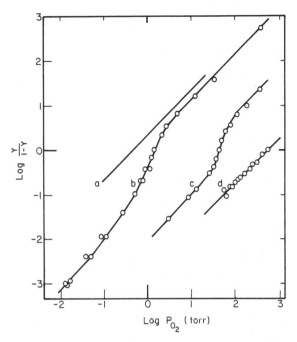

Figure 24. O_2 Binding in solid-state model compounds Cooperative effects a: FeTpivPP(1-MeIm) b: FeTpivPP(2-MeIm) c: FeTpivPP(1,2-Me$_2$Im) d: FeTpivPP(2-MeIm)·EtOH.

of the high- and low-affinity forms, respectively. These are presented in Table 14, along with values for human Hb. In all cases the stoichiometry of O_2 binding was established to be one O_2 molecule per Fe atom. Complete reversibility has been observed with each sample: data were collected at random over the range of O_2 pressures used, and even after

Table 14 Cooperativity in O_2 Binding

System	$P_{1/2}$ (25°C) (torr)		Hill Coefficient	Reference
	High Affinity "R"	Low Affinity "T"		
FeTpivPP(1-MeIm)	0.5	(Noncooperative)	1.0	189
FeTpivPP(2-MeIm)	0.7	14	2.6	189
FeTpivPP(1,2-Me$_2$Im)	14	112	3.0	189
FeTpivPP(2-MeIm)·EtOH	(Noncooperative)	620	1.0	159
Hb A, stripped, pH 7.4	0.3	9	2.5	88

>50 cycles between O_2 and vacuum no discernible change had occurred in the O_2 binding by the sample. The results are also reproducible from sample to sample. In order to demonstrate that these experiments were free of systematic error, a sample of FeTpivPP(1-MeIm) was examined in the same apparatus; this porphyrin had been previously studied (190) with another apparatus and found to be noncooperative in its O_2 binding in the solid state. Such remained the case in our present study: over the range of 20 to 99% oxygenation no deviation from simple noncooperative, Langmuir behavior was observed, and $P_{1/2}$ agreed well with the previous work.

Amazingly FeTpivPP(2-MeIm)·EtOH (in which an ethanol of solvation is hydrogen bonded to the N—H of the 2-MeIm), when oxygenated in the presence of ethanol vapor, binds O_2 very weakly in a *non*cooperative manner (159). This is not an experimental artifact, having been confirmed on two separately prepared samples using both adsorption and desorption measurements. In addition, after being oxygenated under ethanol, a sample of FeTpivPP(2-MeIm)·EtOH can be converted to FeTpivPP(2-MeIm) by evacuation, and the O_2 binding of the desolvated solid regains its high affinity and cooperativity.

Hb shows rather wide variations in its low-affinity (T state) O_2 binding constants, depending on H^+, Cl^-, and phosphate concentrations. Our model systems of course do not have this characteristic, and so comparisons are best drawn between our models and Hb stripped of interfering ions, as shown in Table 14. The similarity between the picket fence porphyrins and Hb A is especially apparent for FeTpivPP(2-MeIm). This is graphically demonstrated in Figure 25: the binding affinities and the degree of cooperative oxygenation of FeTpivPP(2-MeIm) and stripped Hb are of the same order.

The molecular mechanism of cooperativity in these complexes requires further investigation. As observed earlier, we were able to isolate single crystals of FeTpivPP(2-MeIm)·EtOH, and crystal structures were determined for both the deoxy *and* oxy forms (34) (oxygenation of the same crystals did not disrupt their crystallinity). Unfortunately, since the solvated solids are not cooperative in their O_2 binding, one can only use those structures in a general way to speculate on the mechanism of cooperativity. The motion of the 2-MeIm toward the porphyrin upon oxygenation is 0.30 Å in FeTpivPP(2-MeIm)·EtOH, which is ~2% of the dimension of the molecule. Presumably, inasmuch as molecules in the cooperative solid oxygenate, the change in molecular dimensions induces strain in the crystallite. Eventually this strain must be sufficient to induce a conformational change in the solid, which enhances the O_2 affinity of the remaining deoxy sites. This may be analogous to the Hoard–Perutz

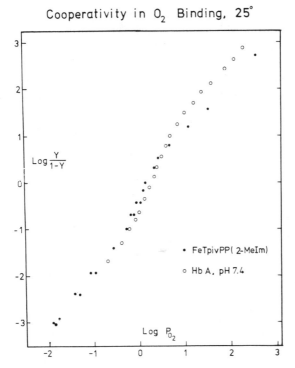

Figure 25. Comparison of Hill plots for FeTpivPP(2-MeIm) in the solid state and HbA in aqueous buffer at pH 7.4.

mechanism of Hb cooperativity. Why this should occur in some complexes and not in others remains unknown.

The decreased affinity of FeTpivPP(1,2-Me$_2$Im) relative to FeTpivPP(2-MeIm) is well explained by the buttressing effect of 1,2-Me$_2$Im. The interaction of the 1-methyl group with the 2-methyl group will increase the steric contact between the 2-methyl group of 1,2-Me$_2$Im and the porphyrin plane relative to 2-MeIm. The slightly greater hindrance is sufficient to decrease further the O$_2$ affinity of the metalloporphyrin. The low affinity observed in FeTpivPP(2-MeIm)·EtOH may be due to the hydrogen-bonded ethanol. As oxygenation occurs, the imidazole in moving toward the porphyrin must drag the ethanol with it, thus making the imidazole movement more difficult.

The similarity in the Hill plots of these complexes and of Hb, as in Figure 25, is dependent on choice of axial base and should not be given exaggerated importance. Furthermore, since these simple complexes have neither protein subunits nor intersubunit contacts, we cannot expect similarities to these portions of the mechanism of Hb cooperativity. However the active sites of Hb and of these solids are similar (both are high-spin, five-coordinate, ferrous porphyrin–imidazole systems capable of binding O_2, CO, etc.), and the effects that cause the cooperativity seen in both systems may be similar *at the metal site*. The differences, protoporphyrinate IX instead of TpivPP, protein subunits instead of crystals, intersubunit contact instead of crystal packing forces, should not of course be ignored. In general the cooperativity of Hb arises from indirect heme–heme interactions, whereas the cooperativity of the "picket fence" porphyrins must arise from more direct interactions. The cooperativity observed in the solid state of these picket fence complexes does demonstrate that O_2 affinities are exquisitely sensitive to axial ligation effects.

Note: Since the initial writing of this chapter, a number of relevant studies have appeared in press. Unfortunately we cannot review these here. As a courtesy to the reader and to the authors of those papers, however, we have added a representative (but undoubtedly incomplete) sampling to our references (202–209).

REFERENCES

1. M. F. Perutz, *Br. Med. Bull.*, **32**, 193, 195 (1976).
2. J. C. Kendrew, R. E. Dickerson, B. E. Strandberg, R. G. Hart, D. R. Davies,
3. D. C. Phillips, and V. C. Shore, *Nature*, **185**, 422 (1960).
4. H. Scouloudi, *J. Mol. Biol.*, **40**, 353 (1969).
5. E. E. Lattman, C. E. Nockolds, R. H. Kretsinger, and W. E. Love, *J. Mol. Biol.*, **60**, 271 (1971).
6. W. A. Hendrickson and W. E. Love, *Nat. New Biol.*, **232**, 197 (1971).
7. E. Antonini and M. Brunori, *Hemoglobin and Myoglobin in their Reactions with Ligands*, American Elsevier, New York, 1971.
8. K. S. Schmitz and J. M. Schurr, *J. Phys. Chem.*, **76**, 534 (1972).
9. H. E. Stanley, R. Bansil, and J. Herzfeld, in *Metal Ions in Biological Systems*, Vol. 7, H. Sigel, Ed., Marcel Dekker, New York, 1978, pp. 312–351.
10. J. V. Kilmartin, *Br. Med. Bull.*, **32**, 209 (1976).
11. R. Benesch and R. E. Benesch, *Nature*, **221**, 618 (1969).
12. J. M. Pratt, in *Techniques and Topics in Bioinorganic Chemistry*, C. A. McAuliffe, Ed., Halstead Press, New York, 1975.
13. C. A. Appleby, *Biochim. Biophys. Acta*, **60**, 226 (1962).
14. C. F. Phelps, E. Antonini, M. Brunori, and G. Kellett, *Biochem. J.*, **129**, 891 (1972).
15. T. Okazaki, R. W. Briehl, J. B. Wittenberg, and B. A. Wittenberg, *Biochim. Biophys. Acta*, **111**, 485, 496, 503 (1965).
16. A. Rossi-Fanelli, E. Antonini, and A. Caputo, *Adv. Protein Chem.*, **19**, 74 (1964).
16a. M. F. Perutz, and H. Lehmann, *Nature*, **219**, 902 (1968).
17. K. M. Smith, *Porphyrins and Metalloporphyrins*, Elsevier, New York, 1975.

REFERENCES

17a. D. Dolphin, *The Porphyrins*, Academic Press, New York, 1978, Vols. 1–7.
18. C. A. Reed, *In Metal Ions in Biological Systems*, Vol. 7, In H. Siegel, Ed., Marcel Dekker, New York, 1978.
19. J. W. Buchler, *Angew. Chem. Int. Ed. Eng.*, **17**, 407 (1978).
20. J. P. Collman, *Acc. Chem. Res.*, **10**, 265 (1977).
21. R. J. P. Williams, *Fed. Proc.*, **20**, 5 (1961).
22. J. L. Hoard, M. J. Hamor, T. A. Hamor, and W. S. Caughey, *J. Am. Chem. Soc.*, **87**, 2312 (1965).
23. J. L. Hoard, in *Hemes and Hemoproteins*, B. Chance, R. W. Estabrook, and T. Yonetani, Eds., Academic Press, New York, 1966, pp. 9–24.
24. M. F. Perutz, *Nature*, **228**, 726 (1970).
25. M. Rougee and D. Brault, *Biochemistry*, **14**, 4100 (1975).
26. D. Brault and M. Rougee, *Biochemistry*, **13**, 4591 (1974).
27. D. Brault and M. Rougee, *Biochem. Biophys. Res. Commun.*, **57**, 654 (1974).
28. J. P. Collman and C. A. Reed, *J. Am. Chem. Soc.*, **95**, 2048 (1973).
29. D. Brault and M. Rougee, *Biochemistry*, **13**, 4598 (1974).
30. J. P. Collman, J. L. Hoard, N. Kim, G. Lang, and C. A. Reed, *J. Am. Chem. Soc.*, **97**, 2676 (1975).
31. H. Goff, G. N. LaMar, and C. A. Reed, *J. Am. Chem. Soc.*, **99**, 3641 (1977).
32. J. L. Hoard, *Ann. N.Y. Acad. Sci.*, **206**, 18 (1973).
33. J. P. Collman, N. Kim, J. L. Hoard, G. Lang, L. J. Radonovich, and C. A. Reed, Abstracts, 167th Meeting ACS, Los Angeles, 1974, INOR 29; D. A. Buckingham, J. P. Collman, J. L. Hoard, G. Lang, L. J. Radanovitch, C. A. Reed, and W. T. Robinson, *J. Am. Chem. Soc.*, to be submitted.
34. G. B. Jameson, F. S. Molinaro, J. A. Ibers, J. P. Collman, J. I. Brauman, E. Rose, and K. Suslick, *J. Am. Chem. Soc.*, **100**, 6769 (1978).
35. J. F. Kirner, C. A. Reed, and W. R. Scheidt, *J. Am. Chem. Soc.*, **99**, 2557 (1977).
36. L. F. Ten Eyck and A. Arnone, *J. Mol. Biol.*, **100**, 3 (1976).
37. T. G. Spiro and J. M. Burke, *J. Am. Chem. Soc.*, **98**, 5482 (1976); J. Kincaid, P. Stein, and T. G. Spiro, *Proc. Natl. Acad. Sci. USA*, **76**, 549 (1979).
38. P. Eisenberger, R. G. Shulman, B. M. Kincaid, G. S. Brown, and S. Ogawa, *Nature*, 274, 30 (1978).
39. H. Goff and G. N. LaMar, *J. Am. Chem. Soc.*, **99**, 6599 (1977).
40. C. K. Chang and T. G. Traylor, *J. Am. Chem. Soc.*, **95**, 5810 (1973).
41. W. S. Brinigar, C. K. Chang, J. Geibel, and T. G. Traylor, *J. Am. Chem. Soc.*, **96**, 5599 (1974).
42. C. K. Chang and T. G. Traylor, *Proc. Natl. Acad. Sci. USA*, **70**, 2647 (1973).
43. M. Momenteau, M. Rougee, and B. Loock, *Eur. J. Biochem.*, **71**, 63 (1976).
44. A. R. Battersby, S. G. Hartley, and M. D. Turnbull, *Tetrahedron Lett.*, **34**, 3109 (1978).
45. T. R. Halbert, Ph.D. Dissertation, Stanford University, 1977.
46. J. P. Collman, J. I. Brauman, K. M. Doxsee, T. R. Halbert, and K. Suslick, *Proc. Natl. Acad. Sci. USA.*, **75**(2), 564 (1978).

47. J. Almog, J. E. Baldwin, and J. Huff, *J. Am. Chem. Soc.*, **97**, 227 (1975).
48. J. E. Baldwin, T. Klose, and M. Peters, *J. Chem. Soc. Chem. Commun.*, 881 (1976).
49. H. Dieckmann, C. K. Chang, and T. G. Traylor, *J. Am. Chem. Soc.*, **93**, 4068 (1971).
50. A. R. Battersby, D. G. Buckley, S. G. Hartley, and M. D. Turnbull, *J. Chem. Soc. Chem. Commun.*, 879 (1976).
51. H. Ogoshi, H. Sugimoto, and Z. I. Yoshida, *Heterocycles*, **3**, 1146 (1975).
52. B. B. Wayland, L. F. Mehne, and J. Swartz, *J. Am. Chem. Soc.*, **100**, 2379 (1978).
53. T. Mincey and T. G. Traylor, *J. Am. Chem. Soc.*, **101**, 765 (1979).
54. L. J. Radonovich, A. Bloom, and J. L. Hoard, *J. Am. Chem. Soc.*, **94**, 2066 (1972).
55. M. E. Kastner, W. R. Scheidt, T. Mashika, and C. A. Reed, *J. Am. Chem. Soc.*, **100**, 666 (1978).
56. D. M. Collins, W. R. Schedit, and J. L. Hoard, *J. Am. Chem. Soc.*, **94**, 6689 (1972).
57. B. Chevrier, Th. Diebold, and R. Weiss, *Inorg. Chim. Acta*, **19**, L57 (1976).
58. A. Warshel, *Proc. Natl. Acad. Sci. USA*, **74**, 1789 (1977).
59. D. B. Olafson and W. A. Goddard, *Proc. Natl. Acad. Sci. USA*, **74**, 1315 (1977).
60. C. A. Reed and W. R. Scheidt, personal communication.
61. W. R. Scheidt, *Acc. Chem. Res.*, **10**, 339 (1977).
62. S. M. Peng and J. A. Ibers, *J. Am. Chem. Soc.*, **98**, 8032 (1976).
63. J. P. Collman, J. I. Brauman, T. R. Halbert, and K. S. Suslick, *Proc. Natl. Acad. Sci. USA*, **73**, 3333 (1976).
64. J. W. Buchler, in *Porphyrins and Metalloporphyrins*, K. M. Smith, Ed., Elsevier, New York, 1975, p. 215.
65. T. Mashiko, M. E. Kastner, K. Spartalian, W. R. Scheidt, and C. A. Reed, *J. Am. Chem. Soc.*, **100**, 6354 (1978).
66. M. Zobrist and G. N. LaMar, *J. Am. Chem. Soc.*, **100**, 1944 (1978).
67. J. L. Hoard, in *Porphyrins and Metalloporphyrins*, K. M. Smith, Ed., Elsevier, New York, 1975, pp. 317–380.
68. M. A. Torrens, D. K. Straub, and L. M. Epstein, *J. Am. Chem. Soc.*, **94**, 4160 (1972).
69. T. H. Moss, H. R. Lilienthal, G. Moleski, G. A. Smythe, M. C. McDaniel, and W. S. Caughey, *J. Chem. Soc. Chem. Commun.*, 263 (1972).
70. A. B. Hoffman, D. M. Collins, V. W. Day, E. B. Fleischer, T. S. Srivastava, and J. L. Hoard, *J. Am. Chem. Soc.*, **94**, 3620 (1972).
71. D. H. Dolphin, J. R. Sams, and T. S. Tsin, *Inorg. Chem.*, **16**, 711 (1977).
72. E. B. Fleischer and D. A. Fine, *Inorg. Chim. Acta*, **29**, 267 (1978).
73. C. L. Coyle, P. A. Rafson, and E. H. Abbot, *Inorg. Chem.*, **12**, 2007 (1973).
74. P. R. Ciaccio, J. V. Ellis, M. E. Munson, G. L. Kedderis, F. X. McConville, and J. M. Duclos, *J. Inorg. Nucl. Chem.*, **38**, 1885 (1976).
75. I. A. Cohen and W. S. Caughey, in *Hemes and Hemoproteins*, B. Chance, R. W. Estabrook, and T. Yonetani, Eds., Academic Press, New York, 1966, p. 577.
76. I. A. Cohen and W. S. Caughey, *Biochemistry*, **7**, 636 (1968).
77. N. Sadisavan, H. I. Eberspaecher, W. H. Fuchsman, and W. S. Caughey, *Biochemistry*, **8**, 534 (1969).
78. O. K. Kao and J. H. Wang, *Biochemistry*, **4**, 342 (1965).

REFERENCES

79. G. S. Hammond and C. H. S. Wu, *Adv. Chem. Ser.*, **77**, 186 (1968).
80. D. H. Chin, J. DelGaudio, G. N. LaMar, and A. L. Balch, *J. Am. Chem. Soc.*, **99**, 5486 (1977).
81. H. B. Dunford and J. S. Stillman, *Coord. Chem. Rev.*, **19**, 187 (1976).
82. M. Cher and N. Davidson, *J. Am. Chem. Soc.*, **77**, 793 (1955).
83. P. George, *J. Chem. Soc.*, 4349 (1954).
84. F. Basolo, B. M. Hoffman, and J. A. Ibers, *Acc. Chem. Res.*, **8**, 384 (1975).
85. B. M. Hoffman and D. H. Petering, *Proc. Natl. Acad. Sci. USA*, **67**, 637 (1970).
86. C. A. Spilburg, B. M. Hoffman, and D. H. Petering, *J. Biol. Chem.*, **247**, 4219 (1972).
87. B. M. Hoffman, C. A. Spilburg, and D. H. Petering, *Cold Springs Harbor Symp. Quant. Biol.*, **36**, 343 (1971).
88. K. Imai, T. Yonetani, and M. Ikeda-Saito, *J. Mol. Biol.*, **109**, 83 (1977).
89. R. S. Drago, T. Beugelsdijk, J. A. Breese, and J. P. Cannady, *J. Am. Chem. Soc.*, **100**, 5374 (1978).
90. J. P. Collman, J. I. Brauman, K. M. Doxsee, T. R. Halbert, S. E. Hayes, and K. S. Suslick, *J. Am. Chem. Soc.*, **100**, 2761 (1978).
91. B. B. Wayland and M. E. Abd-Elmageed, *J. Am. Chem. Soc.*, **96**, 4809 (1974).
92. P. Madura and W. R. Scheidt, *Inorg. Chem.*, **15**, 3182 (1976).
93. J. M. Assour, *J. Chem. Phys.*, **43**, 2477 (1965).
94. F. A. Walker, *J. Am. Chem. Soc.*, **95**, 1150, 1154 (1973).
95. F. A. Walker, *J. Am. Chem. Soc.*, **92**, 4235 (1970).
96. W. R. Scheidt, *J. Am. Chem. Soc.*, **96**, 90 (1974).
97. P. N. Dwyer, P. Madura, and W. R. Scheidt, *J. Am. Chem. Soc.*, **96**, 4815 (1974).
98. W. R. Scheidt and J. L. Hoard, *J. Am. Chem. Soc.*, **95**, 8281 (1973).
99. W. R. Scheidt, *J. Am. Chem. Soc.*, **96**, 84 (1974).
100. R. G. Little and J. A. Ibers, *J. Am. Chem. Soc.*, **96**, 4440 (1974).
101. D. Fremy, *Ann. Chim. Phys.*, **35**, 231 (1852).
102. G. McLendon and A. E. Martell, *Coord. Chem. Rev.*, **19**, 1 (1976).
103. R. S. Gall, J. F. Rogers, W. P. Schaefer, and G. C. Christoph, *J. Am. Chem. Soc.*, **98**, 5135 (1976).
104. C. K. Chang, *J. Chem. Soc. Chem. Commun.*, 800 (1977).
105. H. C. Stynes and J. A. Ibers, *J. Am. Chem. Soc.*, **94**, 1559 (1972).
106. D. V. Stynes, H. C. Stynes, B. R. James, and J. A. Ibers, *J. Am. Chem. Soc.*, **95**, 1796 (1973).
107. H. C. Stynes and J. A. Ibers, *J. Am. Chem. Soc.*, **94**, 5125 (1972).
108. F. S. Molinaro, R. G. Little, and J. A. Ibers, *J. Am. Chem. Soc.*, **99**, 5628 (1977).
109. T. J. Beugelsdijk and R. S. Drago, *J. Am. Chem. Soc.*, **97**, 6466 (1975).
110. B. S. Tovrog, D. J. Kitko and R. S. Drago, *J. Am. Chem. Soc.*, **98**, 5144 (1976).
111. J. A. Lauher and J. E. Lester, *Inorg. Chem.*, **12**, 244 (1973).
112. J. E. Baldwin and J. Huff, *J. Am. Chem. Soc.*, **95**, 8477 (1973).
113. C. K. Chang and T. G. Traylor, *J. Am. Chem. Soc.*, **95**, 8477 (1973).
114. J. P. Collman, R. R. Gagne, C. A. Reed, T. R. Halbert, G. Lang, and W. T. Robinson, *J. Am. Chem. Soc.*, **97**, 1427 (1975).

115. D. L. Anderson, C. J. Weschler, and F. Basolo, *J. Am. Chem. Soc.*, **96**, 5599 (1974).
116. J. Almog, J. E. Baldwin, R. L. Dyer, J. Huff, and C. J. Wilkerson, *J. Am. Chem. Soc.*, **96**, 5600 (1974).
117. W. S. Brinigar and C. K. Chang, *J. Am. Chem. Soc.*, **96**, 5595 (1974).
118. G. C. Wagner and R. J. Kassner, *J. Am. Chem. Soc.*, **96**, 5593 (1974).
119. H. Hartridge and F. J. W. Roughton, *Proc. Roy. Soc.*, **A104**, 395 (1933).
120. M. Brunori and T. M. Schuster, *J. Biol. Chem.*, **244**, 4046 (1969).
121. G. Amiconi, E. Antonini, M. Brunori, H. Fomaneck, and R. Huber, *Eur. J. Biochem.*, **31**, 52 (1972).
122. G. Ilgenfritz and T. M. Schuster, *J. Biol. Chem.*, **249**, 2959 (1974).
123. Q. H. Gibson, *J. Physiol.*, **134**, 112 (1956).
124. R. W. Noble, Q. H. Gibson, M. Brunori, E. Antonini, and J. Wyman, *J. Biol. Chem.*, **244**, 3905 (1969).
125. T. Buecher and E. Negelein, *Biochem. Z.*, **311**, 163 (1941).
126. B. M. Hoffman and Q. H. Gibson, *Proc. Natl. Acad. Sci. USA.*, **75**, 21 (1978).
127. Q. H. Gibson, *J. Biol. Chem.*, **245**, 3285 (1970).
128. T. Imamura, A. Riggs, and Q. H. Gibson, *J. Biol. Chem.*, **247**, 521 (1972).
129. C. J. Weschler, D. L. Anderson, and F. Basolo, *J. Chem. Soc. Chem. Commun.*, 757 (1974).
130. C. J. Weschler, D. L. Anderson, and F. Basolo, *J. Am. Chem. Soc.*, **97**, 6707 (1975).
131. C. K. Chang and T. G. Traylor, *Proc. Natl. Acad. Sci. USA*, **72**, 1166 (1975).
132. J. Geibel, J. Cannon, D. Campbell, and T. G. Traylor, *J. Am. Chem. Soc.*, **100**, 3575 (1978).
133. V. S. Sharma, J. F. Geibel, and H. M. Ranney, *Proc. Natl. Acad. Sci. USA*, **75**, 3747 (1978).
134. J. H. Wang, *J. Am. Chem. Soc.*, **80**, 3168 (1958).
135. J. H. Wang, *Acc. Chem. Res.*, **3**, 90 (1970).
136. C. K. Chang and T. G. Traylor, *Proc. Natl. Acad. Sci. USA*, **70**, 2647 (1973).
137. E. Tsuchida, K. Honda, and H. Sata, *Biopolymers*, **13**, 2147 (1974).
138. E. Tsuchida and H. Honda, *Polym. J.*, **7**, 498 (1975).
139. E. Tsuchida, K. Honda, and H. Sata, *Inorg. Chem.*, **15**, 352 (1976).
140. K. Honda, S. Hata, and E. Tsuchida, *Bull. Chem. Soc. Japan*, **49**, 868 (1976).
141. E. Tsuchida, E. Hasegawa, and K. Honda, *Biochem. Biophys. Acta*, **427**, 520 (1976).
142. E. Tsuchida, E. Hasegawa, and K. Honda, *Biochem. Biophys. Res. Commun.*, **67**, 846 (1975).
143. E. Tsuchida and K. Honda, *Chem. Lett.*, 119 (1975).
144. J. H. Fuhrhop, S. Besecke, W. Vogt, J. Ernst, and J. Subramanian, *Makromol. Chem.*, **178**, 1621 (1977).
145. O. Leal, D. L. Anderson, R. G. Bowman, F. Basolo, and R. L. Burwell, *J. Am. Chem. Soc.*, **97**, 5125 (1975).
146. H. R. Allcock, P. P. Greigger, J. E. Gardner, and J. L. Schmutz, *J. Am. Chem. Soc.*, **101**, 606 (1979).
147. E. Bayer and G. Holzbach, *Angew. Chem. Int. Ed. Eng.*, **16**, 117 (1977).

REFERENCES

148. E. Tsuchida, E. Hasegawa, and T. Kanayama, *Macromolecules*, **11**, 947 (1978).
149. H. Ledon and Y. Brigandat, *J. Orgmet. Chem.*, **165**, C25 (1979).
150. D. Dieckmann, C. K. Chang, and T. G. Traylor, *J. Am. Chem. Soc.*, **93**, 4068 (1971).
151. A. R. Amundsen and C. Vaska, *Inorg. Chim. Acta*, **14**, L49 (1975).
152. L. Vaska, A. R. Amundsen, R. Brady, B. R. Flynn, and H. Nakai, *Finn. Chem. Lett.*, 66 (1974).
153. J. P. Collman, R. R. Gagne, T. R. Halbert, J. C. Marchon, and C. A. Reed, *J. Am. Chem. Soc.*, **95**, 7860 (1973).
154. J. Almog, J. E. Baldwin, R. C. Dyer, and M. Peter, *J. Am. Chem. Soc.*, **97**, 227 (1975).
155. L. K. Gottwald and E. F. Ullman, *Tetrahedron Lett.*, 3071 (1969).
156. C. H. Barlow, J. C. Maxwell, W. J. Wallace, and W. S. Caughey, *Biochem. Biophys. Res. Commun.*, **55**, 91 (1973).
157. J. C. Maxwell, J. A. Volpe, C. H. Barlow, and W. S. Caughey, *Biochem. Biophys. Res. Commun.*, **58**, 166 (1974).
158. J. M. Burke, J. R. Kincaid, S. Peters, R. R. Gagne, J. P. Collman, and T. G. Spiro, *J. Am. Chem. Soc.*, **100**, 6083 (1978).
159. K. S. Suslick, Ph.D. Dissertation, Stanford University, 1978.
160. J. C. Maxwell and W. S. Caughey, *Biochem. Biophys. Res. Commun.*, **960**, 1309 (1974).
161. J. D. Landels and G. A. Rodley, *Synth. Inorg. Met.-Org. Chem.*, **2**, 65 (1972).
162. D. A. White, A. J. Solodar, and M. M. Baizer, *Inorg. Chem.*, **11**, 2160 (1972).
163. S. K. Cheung, C. J. Grimes, J. Wong, and C. A. Reed, *J. Am. Chem. Soc.*, **98**, 5028 (1976).
164. L. Vaska, *Acc. Chem. Res.*, **9**, 175 (1976).
165. G. Lang, *Q. Rev. Biophys.*, **3**, 1 (1970).
166. T. Kent, K. Spartalian, G. Lang, and T. Yonetani, *Biochim. Biophys. Acta*, **490**, 331 (1977).
167. A. Trautwein, H. Eicher, and A. Mayer, *J. Chem. Phys.*, **52**, 2473 (1970).
168. H. Eicher and A. Trautwein, *J. Chem. Phys.*, **50**, 2540 (1969).
169. K. Spartalian, G. Lang, J. P. Collman, R. R. Gagne, and C. A. Reed, *J. Chem. Phys.*, **63**, 5375 (1975).
170. T. A. Kent, K. Spartalian, G. Lang, T. Yonetani, C. A. Reed, and J. P. Collman, *Biochim. Biophys. Acta*, **580**, 245 (1979).
171. M. Cerdonio, A. C. Castellano, F. Mogno, B. Pispisa, G. L. Romani, and S. Vitale, *Proc. Natl. Acad. Sci. USA*, **74**, 398 (1977).
172. M. Cerdonio, A. C. Castellano, L. Calabrese, S. Morante, B. Pispisa, and S. Vitale, *Proc. Natl. Acad. Sci. USA*, **74**, 4916 (1978).
173. L. Pauling and C. D. Coryell, *Proc. Natl. Acad. Sci. USA*, **22**, 210 (1936).
174. L. Pauling, *Proc. Natl. Acad. Sci. USA*, **74**, 2612 (1977).
175. D. F. Evans, *J. Chem. Soc.*, 2005 (1959).
176. G. C. Brackett, P. L. Richards, and W. S. Caughey, *J. Chem. Phys.* **54**, 4383 (1971).
177. P. M. Champion and A. J. Sievers, *J. Chem. Phys.*, **66**, 1819 (1977).
178. J. C. Hoard and W. R. Scheidt, *Proc. Natl. Acad. Sci. USA*, **70**, 3919 (1973).
179. T. Takano, *J. Mol. Biol.*, **110**, 569 (1977).

180. G. Fermi, *J. Mol. Biol.*, **97**, 237 (1975).
181. R. Huber, O. Epp, and H. Formanek, *J. Mol. Biol.*, **52**, 349 (1970).
182. G. B. Jameson, G. A. Rodley, W. T. Robinson, R. R. Gagne, C. A. Reed, and J. P. Collman, *Inorg. Chem.*, **17**, 850 (1978).
183. S. E. V. Phillips, *Nature*, **273**, 247 (1978).
184. E. Weber, W. Steigemann, T. A. Jones, and R. Huber, *J. Mol. Biol.*, **120**, 327 (1978).
185. H. C. Watson and C. L. Nobbs, *Coll. Ges. Biol. Chem.*, **19**, 37 (1968).
186. B. K. Teo and W. K. Li, *Inorg. Chem.*, **15**, 2005 (1976).
187. M. M. Rohmer, A. Dedieu, and A. Veillard, *Theor. Chim. Acta*, **39**, 189 (1975).
188. R. W. Carrell, C. C. Winterborn, and E. A. Rachmilevite, *Br. J. Haematol.*, **30**, 259 (1975).
189. J. P. Collman, J. I. Brauman, E. Rose, and K. S. Suslick, *Proc. Natl. Acad. Sci. USA*, **75**, 1052 (1978).
190. J. P. Collman, J. I. Brauman, and K. S. Suslick, *J. Am. Chem. Soc.*, **97**, 7185 (1975).
191. M. H. Keyes, M. Falley, and R. Lumry, *J. Am. Chem. Soc.*, **93**, 2035 (1971).
192. M. Brunori, R. W. Noble, E. Antonini, and J. Wyman, *J. Biol. Chem.*, **241**, 5230 (1966).
193. I. Tyuma, K. Shimizu, and K. Imai, *Biochem. Biophys. Res. Commun.*, **43**, 423 (1971).
194. Y. Yonetani, H. Tamamoto, and G. V. Woodrow III, *J. Biol. Chem.*, **249**, 682 (1974).
195. M. Ikedo-Saito, H. Yamamoto, K. Imai, F. J. Kayne, and T. Yonetani, *J. Biol. Chem.*, **252**, 620 (1977).
196. R. Oshino, T. Asakura, K. Takio, N. Oshino, and B. Chance, *Eur. J. Biochem.*, **39**, 581 (1973).
197. Q. H. Gibson and M. H. Smith, *Proc. Roy. Soc.*, **B163**, 206 (1965).
198. J. L. Hoard, *Science*, **174**, 1295 (1971).
199. S. Kilmartin, N. L. Anderson, and S. Ogawa, *J. Mol. Biol.*, **123**, 71 (1978).
200. B. R. Gelin and M. Karplus, *Proc. Natl. Acad. Sci. USA*, **74**, 801 (1977).
201. R. G. Little and J. A. Ibers, *J. Am. Chem. Soc.*, **96**, 4452 (1974).
202. M. Momenteau, B. Loock, J. Mispelter, and E. Bisagni, *Nouv. J. Chim.*, **3**, 77 (1979).
203. J. P. Collman, M. Marrocco, P. Denisevich, C. Koval, and F. C. Anson, *J. Electroanal. Chem.*, **101**, 117 (1979).
204. R. D. Jones, D. A. Summerville, and F. Basolo, *Chem. Rev.*, **79**, 139 (1979).
205. C. K. Chang in *Advances in Chemistry*, Vol. 173, R. B. King, Ed., A. C. S., Washington, 1979, p. 162.
206. J. Baldwin and C. Chothia, *J. Mol. Biol.*, **129**, 175 (1979).
207. J. Geibel, J. Cannon, D. Campbell, and T. G. Traylor, *J. Am. Chem. Soc.*, **101**, 3575 (1979).
208. D. A. Case, B. H. Huynh, and M. Karplus, *J. Am. Chem. Soc.*, **101**, 4433 (1979).
209. J. R. Budge, P. E. Ellis, R. D. Jones, J. E. Linard, T. Szymanski, F. Basolo, J. E. Baldwin, and R. L. Dyer, *J. Am. Chem. Soc.*, **101**, 4760, 4762 (1979).

CHAPTER 2
Cytochrome P-450, a Versatile Catalyst in Monooxygenation Reactions

MINOR J. COON and RONALD E. WHITE

Department of Biological Chemistry,
The University of Michigan,
Ann Arbor, Michigan

CONTENTS

1. **INTRODUCTION, 75**

2. **COMPONENTS OF THE LIVER MICROSOMAL MONOOXYGENATION SYSTEM, 78**
   - 2.1 Cytochrome P-450, 78
   - 2.2 NADPH-Cytochrome P-450 Reductase, 83
   - 2.3 Phospholipid, 86
   - 2.4 Cytochrome b_5 and NADH-Cytochrome b_5 Reductase, 87
   - 2.5 Interactions of Components, 87

3. **REGIO- AND STEREOSPECIFICITY OF CYTOCHROME P-450-CATALYZED REACTIONS, 91**

4. **SPECTROSCOPIC BEHAVIOR OF CYTOCHROME P-450, 91**
   - 4.1 Relation of Spectroscopic Behavior to Structure at the Heme Locus, 91
   - 4.2 Electronic Spectroscopy, 92
   - 4.3 Electron Paramagnetic Resonance Spectroscopy, 96
   - 4.4 Mössbauer Spectroscopy, 97
   - 4.5 Circular Dichroism, 98
   - 4.6 Magnetic Circular Dichroism, 100
   - 4.7 Other Types of Spectroscopy, 102

5. **MECHANISM OF OXYGEN ACTIVATION AND INSERTION INTO SUBSTRATES, 103**
   - 5.1 Mechanistic Cycle, 103
   - 5.2 Individual Steps of the Cycle, 104
     - 5.2.1 Binding of Substrate, 105
     - 5.2.2 First Reduction, 105
     - 5.2.3 Binding of Dioxygen, 105
     - 5.2.4 Second Reduction, 106
     - 5.2.5 Splitting of the Oxygen–Oxygen Bond, 107
     - 5.2.6 Hydrogen Abstraction and Radical Recombination, 109
     - 5.2.7 Dissociation of Product, 112
   - 5.3 Rate-Limiting Step, 112
   - 5.4 Hydroperoxide-Supported Hydroxylations, 114

ACKNOWLEDGMENTS, 117

REFERENCES, 117

1 INTRODUCTION

About twenty years ago hepatic microsomes were found to contain a previously unrecognized carbon monoxide-binding pigment (1–3). In the intervening years this pigment has been shown to function as a highly versatile oxygenating catalyst in lipid metabolism and in the detoxification of a variety of foreign substances including drugs, anesthetics, petroleum products, pesticides, and carcinogens. The purpose of this chapter is to summarize our current knowledge of the components of this enzyme system and to indicate the probable mechanism by which it catalyzes the dioxygen-dependent hydroxylation of various substrates. We emphasize the liver microsomal enzyme system because of its physiological importance and our own research interests, but particularly significant studies on this pigment from other sources, such as *Pseudomonas putida* (4), are also described.

Two important contributions which facilitated further research progress with the liver microsomal system were concerned with the function and chemical nature of the carbon monoxide-binding pigment. Estabrook and associates (5, 6) showed that the photochemical action spectrum for reversal of inhibition of O_2-dependent microsomal drug and steroid hydroxylation by CO has a maximum at 450 nm. Since the CO difference spectrum of the reduced pigment also has a maximum at 450 nm, such studies established the metabolic role of this microsomal component. Omura and Sato (7) obtained evidence that the CO-binding pigment, which they called "P-450," was a cytochrome of the *b* type which is reducible by both NADH and NADPH under anaerobic conditions and rapidly reoxidizable in the presence of molecular oxygen. When microsomes were incubated with snake venom or deoxycholate, the pigment was converted to a solubilized form called "P-420" because of the shift in the difference spectrum to 420 nm. However the altered pigment had lost biological activity. The terms "cytochrome P-450" and "cytochrome P-420" now widely used to refer to the native and denatured forms of the enzyme, respectively, do not properly indicate its primary function as an oxygenase. In some instances, however, it does indeed function as a cytochrome, as in electron transfer to cytochrome b_5 (8) or to substrates which serve as terminal acceptors, such as nitro and azo compounds (9), epoxides (10), and amine *N*-oxides (11).

Some of the properties of the hydroxylation system were determined by studies with intact microsomal suspensions, but it was evident that many of the more difficult questions concerning reaction mechanisms and the basis for the remarkably broad substrate specificity could not be answered without isolation and characterization of the enzymes involved.

This laboratory solubilized and resolved the enzyme system about ten years ago (12, 13) and has subsequently been concerned with the purification, properties, and interactions of the components, as well as with related mechanistic studies. Solubilization of hepatic microsomes with deoxycholate in the presence of glycerol and other protective agents, followed by anion exchange chromatography in the presence of detergent, yielded three fractions which were required for reconstitution of hydroxylation activity in reaction mixtures supplemented with NADPH (14). These fractions contained the following components: (a) a soluble form of $P\text{-}450_{LM}$,[1] identified spectrally as the CO complex; (b) a soluble form of NADPH-cytochrome P-450 reductase, a flavoprotein previously called NADPH-cytochrome c reductase when the true electron acceptor was unknown; and (c) a heat-stable, chloroform-soluble factor later shown to be a phospholipid (15). The system reconstituted from these components catalyzes the conversion of substrate (RH) to product (ROH), accompanied by the consumption of molecular oxygen and the oxidation of NADPH, as follows:

$$RH + O_2 + NADPH + H^+ \rightarrow ROH + H_2O + NADP^+ \qquad (1)$$

The expected stoichiometry was demonstrated for this reaction provided that suitable correction was made for the endogenous oxidase activity which leads to hydrogen peroxide formation in either the presence or absence of substrate (16):

$$O_2 + NADPH + H^+ \rightarrow H_2O_2 + NADP^+ \qquad (2)$$

The ability of the enzyme system to generate H_2O_2 is of particular interest in view of the action of hydrogen peroxide and a variety of other oxidizing agents (XOOH) in supporting substrate hydroxylation as shown with microsomal suspensions (17–22) and purified $P\text{-}450_{LM}$ (23), as follows:

$$RH + XOOH \rightarrow ROH + XOH \qquad (3)$$

The scheme shown in Figure 1 illustrates the process of electron transfer from NADPH to the reductase and in turn to $P\text{-}450_{LM}$, resulting in the reduction of one atom of molecular oxygen to H_2O while the other undergoes incorporation into the substrate. Alternatively molecular oxygen may be reduced directly to H_2O_2. Substrate is shown as binding to the

---

The following abbreviations are used: $P\text{-}450_{LM}$, liver microsomal cytochrome P-450; $P\text{-}450_{LM2}$ and $P\text{-}450_{LM4}$, phenobarbital- and 5,6-benzoflavone-inducible forms, respectively, of $P\text{-}450_{LM}$, so designated according to their electrophoretic properties; and dilauroyl-GPC, dilauroylglyceryl-3-phosphorylcholine.

INTRODUCTION

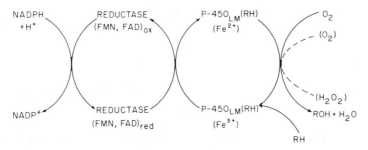

Figure 1. Electron transfer reactions in liver microsomal system leading to cytochrome P-450-catalyzed substrate hydroxylation. RH represents any of a wide variety of substrates which undergo monooxygenation to form the corresponding product, ROH.

oxidized cytochrome prior to reduction of the iron atom to the ferrous state, in accord with evidence (as described below) that substrate greatly enhances the rate of this process.

Cytochrome P-450$_{cam}$, the camphor-hydroxylating enzyme of *Pseudomonas putida* (4), was the first of this class of enzymes to be purified to homogeneity (24). Accordingly the bacterial cytochrome has served as an excellent model for studies with the less readily purified membrane-bound forms of cytochrome P-450. The bacterial system differs in several respects from its microsomal counterpart, as shown by Gunsalus and associates (25). The scheme presented in Figure 2 indicates that NADH is the preferred electron donor and that an iron–sulfur protein, putidaredoxin, is interposed between the reductase and the oxygenase. All three enzymes have been purified to homogeneity and well characterized (26, 27). A further difference between the bacterial and liver microsomal systems is the absence of a phospholipid requirement by the former.

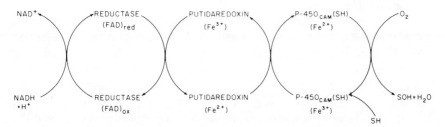

Figure 2. Electron transfer reactions in bacterial system leading to camphor hydroxylation. SH and SOH represent D-camphor and 5-hydroxy-D-camphor, respectively.

2 COMPONENTS OF THE LIVER MICROSOMAL MONOOXYGENATION SYSTEM

2.1 Cytochrome P-450

As reviewed elsewhere (28), the question of whether the many catalytic activities attributed to $P-450_{LM}$ reside in one or more forms of this pigment has been widely investigated; kinetic evidence supported the occurrence of only one form, whereas spectral and genetic evidence, as well as the variable enzyme activities observed in animals treated with different inducing agents, suggested that numerous forms might be involved. More recently, the separation and isolation of some of the individual cytochromes has provided definitive evidence for multiple forms (28). Rabbit liver microsomes apparently contain as many as six forms of $P-450_{LM}$, two of which, phenobarbital-inducible $P-450_{LM_2}$ and 5,6-benzoflavone-inducible $P-450_{LM_4}$, have been obtained in an electrophoretically homogeneous state and characterized in a number of ways (29–32). The nomenclature of these forms is based on numerical designation of the cytochromes according to the general recommendation of the Commission on Biochemical Nomenclature for isozymes (33). The various forms are designated by their behavior upon sodium dodecyl sulfate–polyacrylamide gel electrophoresis, and the individual protein bands are numbered according to decreasing mobility and increasing molecular weight as $P-450_{LM_1}$, $P-450_{LM_2}$, and so on (28, 34). The isolation of the cytochrome from 3-methylcholanthrene-treated rats and rabbits and from phenobarbital-treated rats has been reported by investigators at Hoffmann-La Roche (35, 36) and from phenobarbital-treated and 3-methylcholanthrene-treated rabbits by investigators in Japan (37, 38). Cytochrome P-450 has also been purified from liver microsomes of neonatal rabbits induced by 2,3,7,8-tetrachlorodibenzo-p-dioxin (39) and from rabbit lung microsomes (40, 41), as well as from a number of other sources as recently reviewed by Lu and West (42).

Some of the properties of the forms of cytochrome P-450 that have been characterized most thoroughly, including C- and N-terminal residues and partial amino acid sequence, are given in Table I. The two rabbit liver microsomal enzymes, like many other membrane proteins, have limited solubility in aqueous media and readily aggregate. In contrast, the bacterial cytochrome is a soluble protein having no phospholipid requirement for activity. In immunochemical studies no cross-reactions observable by precipitin band formation were detected between anti-LM_2 serum and $P-450_{LM_4}$, or between anti-LM_4 serum and $P-450_{LM_2}$ (43). Competitive binding studies with radiolabeled cytochromes confirmed that rabbit anti-LM_2

Table 1 Properties of Purified Cytochrome P-450

Property	$P\text{-}450_{LM_2}$	$P\text{-}450_{LM_4}$	$P\text{-}450_{cam}$
Inducing agent	Phenobarbital	5,6-Benzoflavone	Camphor
Minimal molecular weight	49,000	55,000	45,000
Apparent molecular weight	300,000	500,000	45,000
Heme content (per polypeptide chain)	1	1	1
C-Terminal amino acid residue	Arginine	Lysine	Valine
N-Terminal amino acid residue	Methionine	(N-Terminus blocked)	Threonine
Absorption maximum (Soret) without substrate added (nm)	418	394	417
Absorption maximum, CO complex (nm)	451	448	447

does not cross-react with $P\text{-}450_{LM_4}$, but slight cross-reactions were detected by this technique between goat anti-LM_2 and $P\text{-}450_{LM_4}$ and between goat anti-LM_4 and $P\text{-}450_{LM_2}$. These results indicate that the two rabbit cytochromes have significant structural differences. On the other hand, highly purified $P\text{-}450_{LM_2}$ and $P\text{-}450_{cam}$ show immunological cross-reaction by competitive binding and inhibition of catalytic activity and upon treatment with cyanogen bromide yield small heme-containing peptides of highly similar amino acid composition (44).

As shown in the table, the cytochromes vary in the C-terminal (31, 45, 46) and N-terminal regions (45, 47) of the polypeptide chains. $P\text{-}450_{LM_2}$ has an N-terminal methionine residue followed by a highly hydrophobic segment (-Glu-Phe-Ser-Leu-Leu-Leu-Leu-Leu-Ala-Phe-Leu-Ala-Gly-Leu-Leu-Leu-Leu-Leu-Phe-) (47); this region is similar to those of the short-lived hydrophobic amino-terminal precursor segments of certain other proteins (48). In contrast, $P\text{-}450_{LM_4}$ apparently has a blocking group at the N-terminus (47), and the corresponding sequence in $P\text{-}450_{cam}$ is Thr-Thr-Glu-Thr-Ile-Gln-Ser-Asn-Ala-Asn-Leu-Ala-Pro-Leu- (45).

The absorption maxima of the oxidized forms and reduced CO complexes of these cytochromes are summarized in the table, and the complete absorption spectra are presented in Figure 3. Whereas $P\text{-}450_{LM_2}$ has a Soret band at 418 nm typical of the low-spin state, the corresponding band of $P\text{-}450_{LM_4}$ is shifted to the blue at 394 nm and is indicative of the high-spin state (31). This difference is observed most clearly when detergent has been rigorously removed from the latter cytochrome.

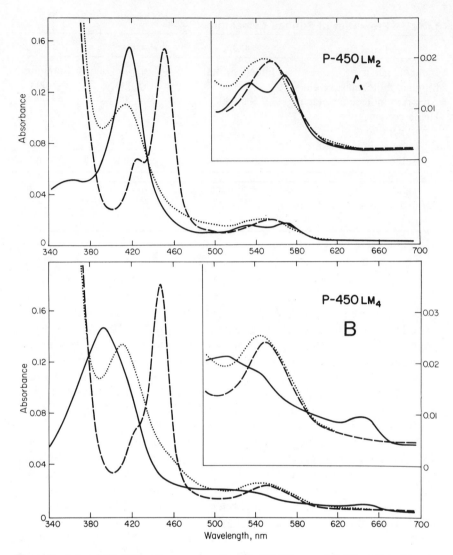

Figure 3. (a) and (b) Absorption spectra of purified P-450$_{LM2}$ and P-450$_{LM4}$ from phenobarbital- and benzoflavone-induced rabbits, respectively, taken from reference 31. (c) Absorption spectra of P-450$_{cam}$ from *Pseudomonas putida*, taken from reference 25. The concentration of the liver cytochromes, based on heme analysis, was 1.40 μM for P-450$_{LM2}$ and 1.53 μM for P-450$_{LM4}$ in 0.1 M potassium phosphate buffer, pH 7.4, containing 20% glycerol. The spectra were recorded at 20 to 22°C: (———) oxidized; (·····) reduced; (- - - -) reduced CO complex. P-450$_{cam}$ (ϵ_{mM} = 87 at 391 nm for oxidized cytochrome with camphor present) was in 0.05 M potassium phosphate buffer, pH 7.4, containing 0.3 mM D-camphor where so indicated. The spectra were recorded at 25°.

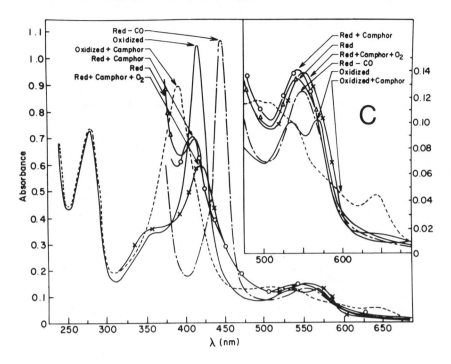

P-450$_{LM_4}$ is converted fairly completely from the 394 nm form to the 418 nm form by the nonionic detergent Renex 690 at a concentration of 0.5% as well as by some impurity present in deionized water or in glass-distilled water stored in plastic containers; this conversion is also favored at higher cytochrome concentrations. Hashimoto and Imai (38) have reported the isolation from 3-methylcholanthrene-treated rabbits of P-450$_{LM_4}$ with this inducer still bound. However since P-450$_{LM_4}$ isolated in this laboratory from benzoflavone-induced, phenobarbital-induced, and uninduced animals has the same spectral properties (31) and furthermore has been exposed to Renex during the purification procedure, it is not clear how bound substrate could account for the spectrum. Oxidized P-450$_{cam}$ has an absorption maximum at 417 nm with camphor absent and at 391 nm with camphor present, and at 447 nm as the reduced CO complex (25). Other species, including the stable oxygen adduct of P-450$_{cam}$ (with camphor present) with a Soret maximum at 418 nm are also shown in Figure 3.

In confirmation of the spin states attributed to the purified cytochromes on the basis of absorption spectra, the EPR spectra of the hepatic (31) and bacterial enzymes (49) are given in Figure 4. The spectrum of P-450$_{LM_2}$ is almost exclusively that of a low-spin ferric hemoprotein, with g values in the high-field region of 1.93, 2.25, and 2.43, which are similar to those reported earlier for a relatively crude solubilized preparation (14) and by Mason et al. (50) for liver microsomes. At higher sensitivity, P-450$_{LM_2}$ shows a small signal at $g = 8.46$, as well as a signal at $g = 4.3$ which we have attributed to traces of an iron impurity. In contrast, the spectrum of P-450$_{LM_4}$ is at least 75% high spin and only partly low spin. The g values are 3.84 and 8.36 in the low-field region and 1.93, 2.26, and 2.42 in the high-field region. EPR spectral evidence for both high- and low-spin forms of cytochrome P-450 in liver microsomal suspensions has been presented by several laboratories (51–53). EPR studies by Tsai et al. (54) and by Griffin et al. (49) have shown that substrate binding causes a spin state change in ferric P-450$_{cam}$. As shown in Figure 4, the substrate-free enzyme exhibits a low-spin signal with g values of 1.92, 2.27, and 2.47 and negligible high-spin content. Addition of camphor converts more than 80% of the hemoprotein to the high-spin form characterized by g values of 7.8, 3.9, and 1.8.

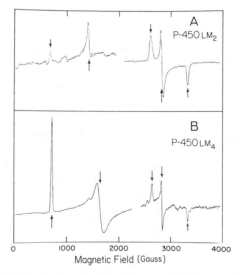

Figure 4. (*a*) and (*b*) EPR spectra of oxidized forms of P-450$_{LM2}$ and LM$_4$, respectively, determined by Dr. J. A. Fee and Dr. R. H. Sands, taken from reference 31. (*c*) EPR spectra of oxidized form of P-450$_{cam}$ with and without camphor present, taken from reference 49.

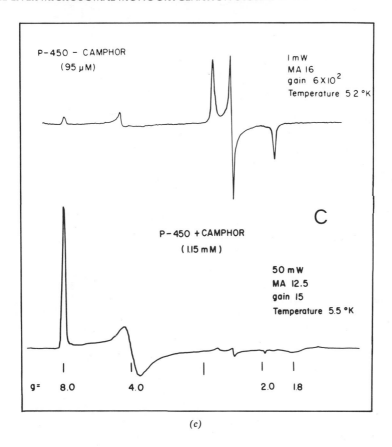

(c)

2.2 NADPH-Cytochrome P-450 Reductase

Several years ago this laboratory reported the extensive purification of NADPH-cytochrome P-450 reductase from detergent-solubilized rat liver microsomes and showed that the ratio of activities toward P-450$_{LM}$ and cytochrome c remained constant throughout the purification procedure (55). NADPH-cytochrome c reductase had previously been characterized as a flavoprotein (56) and purified from lipase(protease)-treated (57) and trypsin-treated microsomes (58). The inactivity of the protease-solubilized enzyme toward cytochrome P-450 (14, 59) is undoubtedly due to cleavage of the protein to produce an altered form which has lost the ability to bind to cytochrome P-450 (60). The purified detergent-solubilized enzyme contains nearly equal amounts of FMN and FAD (55), in accord with the

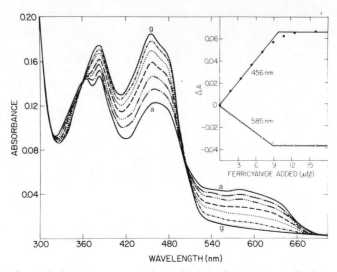

Figure 5. Spectral changes accompanying oxidation of air-stable semiquinone by ferricyanide under aerobic conditions, taken from reference 64. The enzyme (17.4 μM total flavin) was reduced with NADPH (1 mole per mole of flavin) and then allowed to oxidize to the air-stable semiquinone (spectrum a). The other spectra were recorded after the addition of aliquots of ferricyanide solution; spectrum g represents the fully oxidized enzyme, produced upon the addition of a total of 1.1 mole ferricyanide per 2 moles flavin. The inset shows a plot of absorbance changes as a function of ferricyanide added.

report of Iyanagi and Mason (61) that both flavins are present in trypsin-solubilized reductase. Electrophoretically homogeneous preparations of the rat liver reductase with minimal molecular weight of 76,000 (62–64) and rabbit liver reductase with minimal molecular weight of 74,000 (65) have recently been obtained. The preparations from these two sources have highly similar properties.

Whereas studies by Masters et al. (66–68) with reductase solubilized with lipase showed that the enzyme can exist in an air-stable semiquinone state believed to represent a two-electron-reduced form, Iyanagi et al. (69) concluded from studies with the trypsin-solubilized enzyme that this species is actually a one-electron-reduced form. Yasukochi and Masters (62) and Dignam and Strobel (63) have described work with the detergent-solubilized reductase which disagrees with the conclusion of Iyanagi et al. and supports the earlier view. We have recently established that the air-stable semiquinone is a one-electron-reduced form (64), and Masters and her associates[2] now agree with this conclusion. As shown in Figure 5, the air-stable semiquinone (*spectrum a*) has a broad absorption max-

B. S. S. Masters, J. A. Peterson, and Y. Yasukochi, personal communication.

imum at 585 nm with a shoulder at about 630 nm. Upon the addition of 1.1 mole ferricyanide per 2 moles flavin, the fully oxidized enzyme (*spectrum g*) was obtained with maxima at 276, 384, and 456 nm and a shoulder at 485 nm. Since ferricyanide is a one-electron acceptor and the enzyme contains two molecules of flavin, it is concluded that the air-stable form contains only one reducing equivalent per molecule of enzyme.

More recently FMN was selectively removed from the reductase by KBr treatment, and the properties of the modified protein were compared with those of the native enzyme, thereby permitting characterization of the individual flavins (70). As shown in Figure 6, the oxidized, FMN-depleted reductase, which has absorbance maxima at 384 and 455 nm, was reduced by NADPH and then oxidized under aerobic conditions. The resulting semiquinone exhibited maximal absorbance in the long wavelength region at 592 nm, with no shoulder evident at 630 nm, and was further oxidized at an appreciable rate. This experiment showed that the semiquinone produced by air oxidation of the reduced, FMN-depleted reductase is similar to the low potential flavin of the native enzyme both in spectral characteristics and in its relatively high reactivity toward oxygen. These and other experiments established that the high and low po-

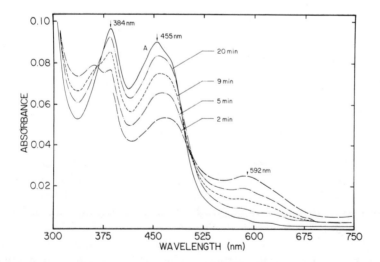

Figure 6. Spectral changes accompanying air oxidation of NADPH reduced, FMN-depleted reductase, taken from reference 70. The FMN-depleted enzyme (8.9 μM total flavin; molar ratio of FMN to FAD, 0.08) was reduced with an excess of NADPH (2.4 moles per mole flavin) under aerobic conditions. Spectrum A is that of the starting oxidized enzyme, and the other spectra shown were recorded at the indicated time periods following the addition of NADPH.

tential flavins of the reductase are FMN and FAD, respectively, and permit the midpoint potentials determined by Iyanagi et al. (69) to be assigned to specific flavins, as follows:

$$\text{FMN} \rightleftharpoons \text{FMNH} \cdot \quad (E_0' = -0.110 \text{ V}) \quad (4)$$

$$\text{FMNH} \cdot \rightleftharpoons \text{FMNH}_2 \quad (E_0' = -0.270 \text{ V}) \quad (5)$$

$$\text{FAD} \rightleftharpoons \text{FADH} \cdot \quad (E_0' = -0.290 \text{ V}) \quad (6)$$

$$\text{FADH} \cdot \rightleftharpoons \text{FADH}_2 \quad (E_0' = -0.365 \text{ V}) \quad (7)$$

Since the FMN-depleted enzyme is capable of accepting electrons from NADPH (as judged by catalysis of ferricyanide reduction) but incapable of transferring electrons to $P\text{-}450_{LM_2}$ (70), the pattern of electron transfer may be NADPH → FAD → FMN → $P\text{-}450_{LM}$. Stopped flow experiments will be required however to verify this hypothesis and establish the detailed manner in which the two individual electron transfer steps to the cytochrome occur. In contrast to our earlier report that $P\text{-}450_{LM}$ is a two-electron acceptor under anaerobic conditions (71), more recent evidence has established that $P\text{-}450_{LM_2}$ and $P\text{-}450_{LM_4}$ consume one electron per hemin molecule and donate one electron to acceptors such as cytochrome b_5 or ferricyanide (8).

2.3 Phospholipid

The third component of the reconstituted liver microsomal enzyme system, phospholipid, strongly stimulates reactions involving $P\text{-}450_{LM}$; these include electron transfer from NADPH via the reductase and substrate hydroxylation, whether supported by molecular oxygen or a peroxy compound (23). Phosphatidylcholine appears to be the most active of the liver microsomal phospholipids tested in the reconstituted system, and certain fatty acyl residues confer more activity than others (15). Various acyl derivatives of glyceryl-3-phosphorylcholine showed, at their optimal concentrations, the following decreasing order of activity with benzphetamine as substrate: dilauroyl, dioleoyl, dipalmitoyl, and a mixture of 1-monopalmitoyl and 1-monostearoyl. The dilauroyl derivative is used routinely in hydroxylation assays because of its high activity and because, unlike the isolated microsomal phosphatidylcholine fraction containing unsaturated fatty acyl residues, it does not readily undergo enzymatic or chemical peroxidation.

The phospholipid causes a decrease in the K_d of benzphetamine for $P\text{-}450_{LM_2}$, as determined from the Type I spectral change, from 6.2×10^{-4} to 1.4×10^{-4} M, and a decrease in the apparent K_d of the reductase, as measured from a frontal boundary gel filtration procedure, from 4.8 × 10^{-7} to 1.2 × 10^{-7} M (60). On the other hand, the phospholipid has no effect on the redox potential of P-450$_{LM_2}$ (61) and does not cause the formation of very large protein aggregates or membranelike structures (72, 73). The effects of phospholipid on the binding of substrate and reductase by P-450$_{LM}$ may explain in part the stimulation of electron transfer from NADPH to the cytochrome but do not rule out other functions of the lipid in substrate hydroxylation, such as at the stage of oxygen insertion.

2.4 Cytochrome b_5 and NADH-Cytochrome b_5 Reductase

Cytochrome b_5 is not an obligatory component of the reconstituted liver microsomal hydroxylation system, since in its absence many substrates examined have turnover numbers equal to, or greater than, those observed with microsomal suspensions (28, 42). On the other hand, although NADH itself supports drug oxidations in hepatic microsomes only poorly, it has a synergistic effect when added along with adequate NADPH (74, 75). From these and related findings it was suggested that cytochrome b_5 functions in the mixed-function oxidation reaction by supplying the second electron to the oxygenated hemoprotein (75). The electron transfer pathways apparently involved are shown in Figure 7. The direct involvement of cytochrome b_5 has been questioned (76, 77), and evidence has been presented that the effect may depend upon the specific form of P-450$_{LM}$ employed (78). Thus much remains to be learned about the possible role of the NADH-dependent pathway in cytochrome P-450-catalyzed oxygenation reactions.

2.5 Interactions of Components

In order to have a better understanding of the nature of the active complex responsible for substrate hydroxylation in the reconstituted liver microsomal system, we have investigated the stoichiometry of interaction of

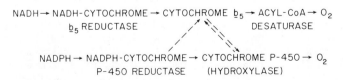

Figure 7. Microsomal electron transfer reactions showing interactions of the NADH- and NADPH-dependent pathways.

the various components with P-450$_{LM_2}$. The extent of benzphetamine binding to the cytochrome was determined by the gel filtration method of Hummel and Dreyer (79) using varying concentrations of the tritiated drug, as shown in Figure 8. The maximal value found for r (which represents moles of substrate bound per mole of P-450$_{LM_2}$ subunit) is 1.1, which indicates that under the conditions employed, including the presence of phosphatidylcholine, the cytochrome has a single binding site for this substrate.

Similar experiments were carried out using ^{14}C-labeled dilauroyl-GPC. As shown in Figure 9, the results indicate that about 20 moles phospholipid are bound per mole P-450$_{LM_2}$ subunit; the apparent K_d is 7.0×10^{-6} M. It should be noted that the critical micellar concentration for this phospholipid, determined under the same conditions according to the method of Bonsen et al. (80), is considerably higher (about 4.5×10^{-5} M).

Evidence for the interaction of NADPH-cytochrome P-450 reductase with P-450$_{LM}$ in the presence of phospholipid and substrate (benzphetamine) is presented in Figure 10. In the experiment shown in the upper part of the figure, a mixture of phospholipid, reductase, benzphetamine, and

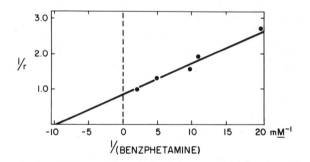

Figure 8. Extent of benzphetamine binding to cytochrome P-450 determined by gel filtration, taken from reference 60. P-450$_{LM2}$ (56 nmole; 5.2 mg protein) in 1.5 ml 0.05 M Tris-acetate buffer, pH 7.5, containing 0.83 mM dilauroyl-GPC, 0.1 mM EDTA, 10% glycerol, and d-[N-methyl-^{3}H]benzphetamine (0.54 μCi/μmole; at the same concentrations as in the equilibrated columns) was applied at 23°C to a Bio-Gel P-6 column, 0.6 × 60 cm, which had been equilibrated with the same buffer solution containing radioactive benzphetamine at the particular concentrations shown in the figure. Each column was eluted with the same solution as used for equilibration, and 1.2 ml fractions were collected and counted. The concentration of free substrate was determined from the baseline level of radioactivity, and that of the bound substrate was determined from the radioactivity in the protein peak corrected for the baseline level. The reciprocal of r (where r is defined as moles substrate bound per mole P-450$_{LM2}$ subunit) is plotted against the reciprocal of the free benzphetamine concentration.

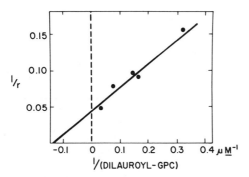

Figure 9. Extent of phosphatidylcholine binding to cytochrome P-450 determined by gel filtration, taken from reference 60. Experiments similar to those in Figure 8 were carried out with P-450$_{LM_2}$ (11.8 nmole; 1.1 mg protein) using radioactive dilauroyl-GPC (0.13 μCi/μmole, having 1-^{14}C-labeled lauroyl residues) and with substrate absent.

P-450$_{LM_2}$ was submitted to gel filtration. The phospholipid, which was at a level which exceeded the critical micellar concentration, appeared first from the column. The cytochrome and the reductase were eluted together in a broad peak, apparently as a complex having a molecular weight of about 500,000, and additional reductase was eluted subsequently in a peak with somewhat lower apparent molecular weight. Hydroxylation activity, measured by the demethylation of benzphetamine upon the addition of NADPH, corresponded closely to the concentration of P-450$_{LM_2}$. In the experiment shown in the lower part of the figure, highly purified, protease-solubilized reductase was used in place of the purified, detergent-solubilized reductase and showed no interaction with the cytochrome. These results clearly indicate the inability of the bromelain-treated reductase to form a complex with P-450$_{LM_2}$ and correlate well with the failure of such reductase preparations to bring about substrate hydroxylation under similar conditions in the reconstituted system (59). Apparently, proteolytic solubilization by bromelain (81) or steapsin (82) removes part of the polypeptide chain which is necessary for binding to cytochrome P-450.

In other experiments evidence has been obtained that reductase and P-450$_{LM_2}$ bind in an equimolar ratio (based on subunit concentrations), as judged by steady-state kinetics, spectral titrations, and CD spectrophotometry (83). In summary, it appears that the stoichiometric ratios for binding in a functional complex are 1 benzphetamine : 1 reductase : about 20 dilauroyl-GPC : 1 P-450$_{LM_2}$.

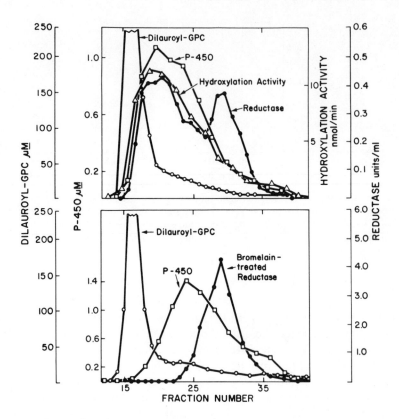

Figure 10. Interaction of cytochrome P-450 with reductase as shown by gel filtration, taken from reference 60. In the experiment shown in the upper part of the figure, a mixture of P-450$_{LM2}$ (25 nmole; 2.7 mg protein), NADPH-cytochrome P-450 reductase (specific activity, 16.7; 0.72 mg protein), radioactive dilauroyl-GPC (1.55 μmole; 0.2 μCi), and benzphetamine (2.2 μmole) in 2.2 ml final volume of 0.05 M Tris buffer, pH 7.5, containing 15% glycerol, was allowed to stand at 23° for 10 min and was then applied to a column of Bio-Gel A-1.5 m, 1.0 × 50 cm, previously equilibrated with Tris buffer, pH 7.5, containing 10 mM MgCl$_2$ and 1.0 mM benzphetamine. The column was eluted with the equilibrating buffer, and 1.3 ml fractions were collected and analyzed for cytochrome P-450 by the CO difference spectrum, the reductase by cytochrome c reduction at 30°, phospholipid by radioactivity, and hydroxylation activity at 30° (after the addition of NADPH) by formaldehyde formation from benzphetamine. In the experiment shown in the lower part of the figure, highly purified bromelain-solubilized NADPH-cytochrome c reductase (specific activity, 57; 0.70 mg protein) was substituted for the usual cytochrome P-450 reductase.

3 REGIO- AND STEREOSPECIFICITY OF CYTOCHROME P-450-CATALYZED REACTIONS

The picture which has emerged from the study of substrate specificity of the various forms of rabbit liver microsomal cytochrome P-450 is that these enzymes have somewhat different but overlapping activities. Despite the great variety of types of reactions catalyzed (hydroxylation, epoxidation, dehalogenation, etc.) and the diversity of substrates attacked, the transformations in many cases show positional selectivity and even stereospecificity (84). For example, P-450$_{LM_2}$ catalyzes the hydroxylation of biphenyl primarily in the 4-position, whereas the other forms tested do not distinguish between the 2- and 4-positions, and P-450$_{LM_2}$ catalyzes the hydroxylation of testosterone primarily in the 16α-position and a mixture of P-450$_{LM_1}$ and Lm$_7$ in the 6β-position (28). More recently we have demonstrated differences in the hydroxylation of R and S warfarin (85) and benzo[a]pyrene and the (−)-*trans*-7,8-diol of benzo[a]pyrene (86) by the various cytochromes. The substrate specificity and regio- and stereoselectivity of the different forms of P-450$_{LM}$ may regulate the balance between activation and detoxification pathways of benzo[a]pyrene and therefore determine the susceptibility of individual tissues, strains, and species to the carcinogenic action of this polycyclic hydrocarbon.

4 SPECTROSCOPIC BEHAVIOR OF CYTOCHROME P-450

4.1 Relation of Spectroscopic Behavior to Structure at the Heme Locus

The chromophore usually probed by a particular spectroscopic technique in the case of cytochrome P-450 is the heme, especially the iron atom coordinated therein. In many cases the behavior of the chromophore may be correlated with molecular structure, geometry, electronic configuration, and oxidation state in the immediate vicinity of the iron. Usually several types of measurements must be considered together to assign an element of structure unambiguously, and heavy reliance has been placed on comparisons to other hemoproteins and to low molecular weight iron porphyrin complexes whose structures are known with varying degrees of certainty. From such studies, using many different forms of spectroscopy, a conceptual model of the structure and function of P-450 has arisen. Here we focus on the structural aspects of the model, while the same elements are considered in a more dynamic framework in Section 5.

The active site of P-450$_{LM}$ contains an iron protoporphyrin IX moiety in a large, relatively open, hydrophobic cleft or depression in the surface of the apoprotein (Fig. 11). The heme is bound, apparently somewhat loosely, by a combination of hydrophobic forces, coulombic attractions, and one or two coordinate-covalent bonds to the central metal ion. The iron is always penta- or hexacoordinate, four of the ligands being contributed by the planar, tetradentate porphyrin ring. The fifth ligand appears to be the same in most if not all states of the cytochrome, provided denaturation has not occurred. It is thought to be a thiolate anion contributed by a cysteine residue of the polypeptide chain. The sixth coordination position of the iron is occupied by an exchangeable ligand, perhaps water, in the native, substrate-free, ferric state. Upon reduction the sixth position becomes the site of dioxygen binding. The other diatomic ligands—carbon monoxide, nitric oxide, and cyanide—alternately may occupy the sixth position under appropriate conditions. Hydrophobic substrates are bound to the surface of the protein and/or the heme on the side of the sixth coordination position. The establishment of this model of the P-450 active site has been aided by the unusual spectral properties exhibited by many of the stable states of the protein. These unusual features are elaborated in the next sections.

4.2 Electronic Spectroscopy

The absorbance spectra exhibited by various states of P-450 in the region 300 to 700 nm are compiled in Table 2. The P-450 hemoproteins from all sources are remarkably similar in spectral properties, making it possible to use the forms somewhat interchangeably for spectral measurements. The substrate-free, ferric, or "resting" state of the cytochrome is fairly

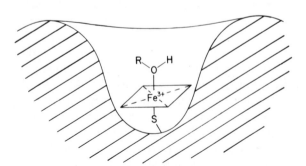

Figure 11. Model of the active site of P-450$_{LM2}$.

Table 2 Electronic Spectra of Cytochrome P-450 States

Cytochrome	Oxidation State	Ligands	Absorbance Maxima (nm)				Spin State	Reference
			n.u.v.[a]	Soret	β	α		
1. $P-450_{LM_2}$	Ferric	Native	360	418	535	568	Low	31
2. $P-450_{LM_4}$	Ferric	Native	—[b]	394	—	645	High	31
3. $P-450_{LM_2}$	Ferrous	Native	—	413	—	544	High	31
4. $P-450_{LM_2}$	Ferrous	Native, CO	370	451	—	552	Low	71
5. $P-450_{cam}$	Ferrous	Native, NO	366	438	—	558	Low	87
6. $P-450_{cam}$	Ferrous	Native, O_2	356	418	555	580s	Low	88

[a] Near ultraviolet.
[b] Dash indicates the absence of a band in that region.

typical of low-spin ferric hemoproteins, with a Soret band at 418 nm and well-defined α- or β-absorbances. The hexacoordinate 418-nm form is reversibly converted to the pentacoordinate, high-spin state upon the binding of substrate, and this change is accompanied by a shift of the Soret transition to higher energy (from 418 nm to 394 nm). The distinct bands in the 500 to 600 nm range disappear in the high-spin form, but a new low-intensity band appears at 645 nm. This latter absorbance is attributed to a charge-transfer transition and is frequently observed in hemoproteins. The form of the cytochrome designated $P-450_{LM_4}$ exists in the high-spin state as isolated, unlike the other forms of P-450 that have been purified. However all forms give nearly the same spectrum when substrate is present.

When P-450 is reduced, either chemically or enzymatically, in the absence of oxygen or other potential ligands, the Soret band loses intensity and moves to 413 nm, and the α,β-bands collapse to a single broad band at 544 nm. This behavior appears to be unique among hemoproteins, all other showing a more intense, longer wavelength peak in the ferrous state. However the most striking difference between P-450 and other hemoproteins is seen when the ferrous form is exposed to carbon monoxide. The Soret band of ferrous carbonyl $P-450_{LM_2}$ is located at an unusually high wavelength, 451 nm. The range of this transition is about 447 to 451 nm for the entire series of isolated P-450 hemoproteins, and the appearance of a band in this range upon reduction of a hemoprotein in the presence of CO serves as the operational definition of cytochrome P-450. In fact, when the enzyme is exposed to denaturing conditions (e.g., high temperature, chaotropic agents, ionic detergents), the Soret band of the ferrous carbonyl form shifts from the 450 nm region to the 420 nm region (e.g., 423 nm for P-450$_{LM_2}$). The 420 nm transition has roughly twice the molar absorptivity as compared to the 450 nm transition, so that the presence of even small amounts of P-420, the denatured form, is readily detectable. The spectrum of P-420 is very similar to that given by non-P-450, CO-binding hemoproteins, such as hemoglobin.

The unprecedented and unexpected long wavelength absorption of ferrous carbonyl P-450 led to a concerted effort by various investigators to reproduce this behavior in well-defined model systems. The first reported success was by Stern and Peisach (89), who were able to elicit such behavior from a carbonyl ferroheme in which propanethiolate was apparently the sixth ligand. The inference was that the unusual Soret band position in ferrous carbonyl P-450 was a direct result of thiolate ligation to the heme and was in agreement with an earlier proposal of such a ligand based on EPR results (90). Subsequent work by many groups (91–98), using systems in which the coordination sphere was better characterized, provided persuasive evidence for the presence of a thiolate at the fifth position in P-450. Table 3 shows spectral characteristics of some of these model iron porphyrin complexes. Line 8 gives the properties of an *n*-butyl thiolate-coordinated ferrous carbonyl porphyrin. The spectrum is very similar to that of ferrous carbonyl P-450. On the other hand, similar complexes in which imidazole, a dialkyl ether, a thioether, or a thiol is the ligand trans to CO give spectra of the normal type, with the Soret band centered around 420 nm (cf. lines 9 to 12). Thus this state of P-450 is modeled well only when thiolate is the fifth ligand. Similarly the ferrous nitrosyl P-450 spectrum is atypical among hemoproteins but is mimicked well by the corresponding thiolate iron prophyrin model (line 13). Consequently we may state that, to the degree to which the structures of the model compounds are known, we are confident that the fifth ligand in ferrous carbonyl and ferrous nitrosyl P-450 is thiolate. These empirical correlations have been given a theoretical foundation by Hanson et al. (101), who invoked the concept of a "hyperporphyrin" spectrum to interpret the spectrum of ferrous carbonyl P-450. They demonstrated with iterative extended Hückel (IEH) calculations that a transition involving charge transfer from a thiolate p-orbital to a porphyrin π^*-orbital can interact strongly with porphyrin $\pi \rightarrow \pi^*$ transitions, resulting in a "split" Soret band, one limb occurring at higher energy, the other at lower energy than the normal Soret band. Other ligands, including thiol, did not have the correct lone-pair orbital energies to allow this orbital mixing to occur and were predicted to not exhibit hyperporphyrin spectra.

Given the probability of a thiolate as fifth ligand, the other stable states of P-450 listed in Table 2 were successfuly modeled, with the exception of the ferrous dioxygen complex. Thus the pentacoordinate high-spin fer-

Table 3   Electronic Spectra of Iron Porphyrin Complexes[a]

Complex	Oxidation State	Ligands	n.u.v.[b]	Soret	β	α	Spin State	Reference
1. FePPIXDME	Ferric	n-BuS$^-$, EtOH	*[c]	416	534	559	Low	99
2. FePPIXDME	Ferric	O$_2$NC$_6$H$_4$S$^-$, 2-MeTHF	*	418	535	566	Low	98
3. FePPIXDME	Ferric	PhCH$_2$S$^-$, Im	*	428	538	568	Low	94
4. FePPIXDME	Ferric	n-BuS$^-$, (CH$_2$)$_5$S	—[d]	432	533	578	Low	99
5. FePPIXDME	Ferric	n-BuS$^-$, nBuS$^-$	377	475	—	565	Low	94
6. FePPIXDME	Ferric	ClC$_6$H$_4$S$^-$, none	—	392	527	645	High	98
7. FePPIX	Ferrous	Ch$_3$S$^-$, none	—	410	—	550	High	92
8. FePPIXDME	Ferrous	n-BuS$^-$, CO	370	451	—	554	Low	100
9. FeTpivPP	Ferrous	N-MeIm, CO	*	427	*	*	Low	91
10. FeTpivPP	Ferrous	THF, CO	*	417	*	*	Low	91
11. FeTpivPP	Ferrous	THT, CO	*	428	*	*	Low	91
12. FeTpivPP	Ferrous	n-PrSH, CO	*	422	*	*	Low	91
13. FePPIX	Ferrous	CH$_3$S$^-$, NO	370	437	—	557	Low	92

[a] FePPIXDME = iron protoporphyrin IX dimethyl ester; FePPIX = iron protoporphyrin IX; FeTpivPP = iron tetra(pivalamidophenyl)porphyrin; 2-MeTHF = 2-methyltetrahydrofuran; Im = imidazole; N-MeIm = N-methylimidazole; THF = tetrahydrofuran; THT = tetrahydrothiophene.
[b] Near ultraviolet.
[c] Asterisk indicates that presence or absence of band was not reported.
[d] Dash indicates the absence of a band in that region.

ric and ferrous P-450 states also appear to have thiolate ligation (see Table 3, lines 6 and 7). Furthermore if we accept the presence of thiolate in the ferric, low-spin P-450, the compounds on lines 1 to 5 of Table 3 allow us to note that the spectrum of the native, low-spin cytochrome is best fit by a hexacoordinate ferric porphyrin coordinated to a thiolate and to an oxygen ligand such as a hydroxyl group or a dialkyl ether. Poor fits are obtained from ligation of an imidazole, a thioether, or a thiol. In fact, the last compound gives another hyperporphyrin spectrum (94). Although an attempt has been made to prepare a thiolate-ligated ferrodioxygen porphyrin (102), the species obtained probably was a ferriporphyrin bisthiolate complex (94). The successful modeling of the dioxygen complex may require the use of the so-called "picket-fence" porphyrins (103).

4.3 Electron Paramagnetic Resonance Spectroscopy

As alluded to earlier, the unusual character of the EPR spectrum, which is diagnostic of a highly rhombic field, suggested an unusual ligation mode in P-450 (90). Only the ferric low- and high-spin states exhibit detectable EPR signals. For the $P-450_{LM}$ these two spectra are presented in Figure 4, and the appropriate g values are given in Section 2.1. The unique nature of the EPR spectrum of P-450 is another criterion for "fitting" of model heme systems to the behavior of the cytochrome. The ferric, high-spin form is not distinctly different from other hemoproteins, but, as seen in Table 4 (lines 1 to 3), thiolate makes a better match than either ethoxide or acetate as the ligand in pentacoordinate ferric porphyrin complexes. More striking evidence for thiolate coordination is provided by the series of hexacoordinate low-spin complexes (lines 4 to 12). In those cases where thiolate is the fifth ligand (i.e., lines 4 to 9), a good fit of the g values to those of P-450 is observed. However when thiolate is not present, as in the compounds on lines 10 to 12, the spectra are distinctly different. Thus the EPR spectrum of P-450 provides further evidence, by reference to model compounds, of thiolate ligation, at least in the ferric states. Apparently the highly polarizable thiolate ligand predominates over the sixth ligand in its effect on the iron d electrons, but the presence of large rhombicity in the nominally axially symmetrical heme bisthiolate complex (line 9) is puzzling. Perhaps one ligand is thiolate and the other is thiol. Thus, although the nature of the sixth ligand in these complexes varies from dialkyl ether to primary amine, large changes in g values are not observed as long as thiolate is the fifth ligand. This damping effect unfortunately prevents a clear assignment of the ligand trans to thiolate in P-450 based on EPR g values alone.

SPECTROSCOPIC BEHAVIOR OF CYTOCHROME P-450

Table 4 EPR Spectra of Ferric Iron Porphyrin Complexes[a]

Complex	Ligands	g Values			Spin State	Reference
1. FePPIXDME	$ClC_6H_4S^-$, none	1.9	4.8	7.2	High	98
2. FePPIXDME	EtO^-, none	1.9	5.5	6.5	High	98
3. FePPIXDME	OAc^-, none	2.0	5.9	—	High	98
4. FeTPP	PhS^-, THF	1.96	2.25	2.34	Low	96
5. FeTPP	PhS^-, acetone	1.94	2.26	2.37	Low	96
6. FeTPP	PhS^-, N-MeIm	1.93	2.26	2.39	Low	96
7. FeTPP	PhS^-, $MeNH_2$	1.96	2.22	2.38	Low	96
8. FeTPP	PhS^-, THT	1.95	2.27	2.36	Low	96
9. FePPIXDME	n-BuS^-, n-BuS^-	1.96	2.23	2.31	Low	94
10. FePPIXDME	N-MeIm, N-MeIm	1.57	2.29	2.90	Low	98
11. FePPIXDME	N-MeIm, EtO^-	1.92	2.15	2.44	Low	98
12. Cytochrome c	Met, His	1.24	2.24	3.06	Low	104

[a] Abbreviations used are as in Table 3; in addition, Met = methionine; His = histidine.

4.4 Mössbauer Spectroscopy

Mössbauer spectroscopy has been performed only on $P-450_{cam}$ because of the relative ease of substitution of the iron by growing the bacteria in a ^{57}Fe-enriched medium. Data from these experiments are compiled in Table 5. Unlike EPR, Mössbauer spectroscopy allows the iron to be

Table 5 Mössbauer Data for Cytochrome P-450 States[a]

Oxidation State	Added Ligand	Spin State	Quadrupole Splitting ΔE_q (mm/sec)	Isomer Shift δ (mm/sec)
Ferric[b]	None	Low	2.75	0.31
Ferric[c]	Camphor	High	0.78	0.34
Ferrous[d]	None	High	2.38	0.77
Ferrous[e]	Camphor	High	2.36	0.76
Ferrous[f]	Camphor, O_2	Low	2.07	0.27
Ferrous[f]	Camphor, CO	Low	0.34	0.25

[a] Cytochrome $P-450_{cam}$ (ref. 105).
[b] 206°K.
[c] 210°K.
[d] 190°K.
[e] 213°K.
[f] 200°K.

probed in all oxidation and ligation states. The ferric low-spin, ferric high-spin, and ferrous high-spin states each give distinctive spectra. The addition of camphor to the ferrous form does not alter the spectrum, a result anticipated from the lack of substrate perturbation of the electronic absorption spectrum. The ferrous dioxygen and ferrous carbonyl spectra are quite different from those of the other states but are not unusual with respect to the corresponding hemoglobin derivatives (see Table 6). Deoxyhemoglobin in fact appears to have a nuclear environment quite similar to that of ferrous P-450. A thiolate-coordinated, high-spin ferriheme model compound reproduces the Mössbauer parameters of substrate-bound ferric P-450 rather well, but other model hemes lacking thiolate show somewhat the same behavior. Thus the most that can be said of the Mössbauer results so far is that they agree with the hypothesis of thiolate ligation in P-450 but do not rule out other ligation modes.

4.5 Circular Dichroism

Circular dichroic spectra have been reported for solubilized $P-450_{LM}$ (71, 110) and for $P-450_{cam}$ (111) but not for the isolated liver microsomal forms

Table 6 Mössbauer Data for Iron Porphyrin Complexes

Complex	Oxidation State	Spin State	Quadrupole Splitting ΔE_q (mm/sec)	Isomer Shift δ (mm/sec)	Reference
FePPIXDME $(O_2NC_6H_4S^-)$[a]	Ferric	High	0.76	0.33	98
FePPIXDME (EtO^-)[a]	Ferric	High	0.46	0.32	98
FePPIX(Im)Cl^-)[a]	Ferric	High	0.78	0.26	106
Acid-metMb[b,c]	Ferric	High	1.3	0.4	107
MetHb(MeSH)[b,d]	Ferric	Low	1.35	0.4	108
Hb[e,f]	Ferrous	High	2.36	0.59	109
$Hb(O_2)$[e,f]	Ferrous	Low	2.25	−0.08	109
Hb(CO)[e,f]	Ferrous	Low	0.34	−0.07	109

[a] 4.2°K.
[b] 195°K.
[c] Acid-metmyoglobin.
[d] Methemoglobin.
[e] Hemoglobin.
[f] 5° K.

P-450$_{LM_2}$ and P-450$_{LM_4}$. This is unfortunate because of the various physical techniques for probing the active site of P-450, CD is the most sensitive to the asymmetry of the environment of the heme chromophore. Nonetheless, the data in Table 7 show that the protein structure around the heme is significantly different in the enzymes from the two sources. None of the CD bands of P-450$_{LM}$ corresponds closely to the absorption maxima in any of the states. Of course the interpretation is complicated by the presence of several forms of P-450$_{LM}$ in the crude, solubilized preparation. However it appears that the Soret band in all states of P-450$_{LM}$, including the ferrous carbonyl derivative, is composed of two or more electronic transitions which are not equally optically active. In contrast, the Soret band of ferrous carbonyl P-450$_{cam}$ exhibits positive ellipticity at 436 nm, nearly equally intense negative ellipticity at 455 nm, and a Cotton effect at 447 nm (corresponding to the absorbance maximum). Thus this band is well described as the sum of two overlapping elements showing opposite optical activity. In the ferrous carbonyl derivatives of both P-450$_{LM}$ and P-450$_{cam}$ a strong, negative CD band is seen at 350 to 360 nm, probably representing the higher energy component of the hyperporphyrin spectrum assigned to this state.

Table 7 Circular Dichroism of Cytochrome P-450

Cytochrome	Oxidation State	Added Ligands	Wavelength (nm)	Ellipticity θ ($10^4 \cdot \deg \cdot cm^2/dmol$)	Reference
1. P-450$_{LM}$	Ferric	Substrate	356	−3.3	71
			425	−17	
2. P-450$_{LM}$	Ferrous	Substrate	395	−7.2	71
3. P-450$_{LM}$	Ferrous	CO	352	−5.1	71
			457	−7.1	
4. P-450$_{cam}$	Ferric	None	350	−6.3	111
			410	−10.6	
5. P-450$_{cam}$	Ferric	Camphor	388	−11.1	111
			507	0	
			539	+2.1	
6. P-450$_{cam}$	Ferrous	None	388	−7.0	111
7. P-450$_{cam}$	Ferrous	CO	360	−6.6	111
			388	0	
			436	+5.0	
			447	0	
			455	−4.2	

4.6 Magnetic Circular Dichroism

Both P-450$_{LM_2}$ and P-450$_{LM_4}$ have been well studied by magnetic circular dichroism (MCD) (38, 112–114). The collected data are presented in Table 8. Since MCD does not require an asymmetric molecule, the spectra of three pertinent model compounds have been reported also, as presented in Table 9. These spectra have not been interpreted in theoretical terms but have been used in an empirical fashion to relate the structure of model hemes to that at the P-450 active site. In agreement with results from electronic spectra and from EPR, only the model iron porphyrin with thiolate ligation (Table 9, line 1) gives a good agreement with the substrate-bound ferric cytochrome. In fact, the agreement is striking. On the other hand, when phenolate (Table 9, line 2), ethoxide (115), or acetate (115)

Table 8 Magnetic Circular Dichroism of Cytochrome P-450

Cytochrome	Oxidation State	Added Ligand	Wavelength (nm)	Molar Ellipticity per Gauss $[\theta]_M$	Reference
1. P-450$_{LM_2}$	Ferric	None	359	−3.9	112
			402	+11.0	
			416	0	
			426	−12.3	
			512	+2.1	
			534	−1.0	
			553	+4.0	
			570	−8.3	
2. P-450$_{LM_4}$	Ferric	Substrate	394	−9.7	38
			407	0	
			414	+3.7	
			550	−4.0	
			647	−2.4	
3. P-450$_{LM_2}$	Ferrous	None	404	+4.9	112
			413	0	
			422	−10.8	
			439	+6.2	
			550	+4.2	
			575	−2.8	
4. P-450$_{LM_2}$	Ferrous	CO	376	−10.0	113
			442	+20.9	
			452	0	
			459	−15.8	
			531	+2.1	
			585	−4.3	

Table 9 Magnetic Circular Dichroism of Iron Porphyrin Complexes

Complex	Oxidation State	Ligands	Wavelength (nm)	Molar Ellipticity per Gauss $[\theta]_M$	Reference
1. FePPIXDME	Ferric	$O_2NC_6H_4S^-$, none	397	−6.8	115
			413	0	
			421	+1.7	
			495	+2.7	
			554	−2.9	
			662	−1.3	
2. FePPIXDME	Ferric	$O_2NC_6H_4O^-$, none	404	+7.0	115
			414	0	
			422	−4.1	
			545	−2.7	
			607	+1.7	
			638	−2.9	
3. FePPIXDME	Ferrous	CH_3S^-, CO	373	−8.2	113
			442	+15.9	
			452	0	
			459	−16.0	
			540	+1.5	
			587	−3.2	
4. FePPIXDME	Ferrous	N-MeIm, CO	411	+24.8	113
			419	0	
			427	−25.3	
			583	−10.4	

is the ligand, the MCD spectrum is much different. Furthermore the MCD spectrum of acid metmyoglobin, which has imidazole and water ligands, shows no similarity either (115).

The most conspicuous feature of the MCD spectra of the thiolate-ligated complexes and of substrate-bound P-450$_{LM_4}$ is the strong *negative* ellipticity in the Soret region. In contrast, the substitution of oxygen or nitrogen ligands for thiolate in the high-spin complexes results in an MCD spectrum with strong *positive* ellipticity in this region. Interestingly the charge-transfer band at 645 nm in the absorption spectrum of P-450$_{LM_4}$ has a corresponding MCD band at a coincident wavelength, indicating that a single transition comprises this absorption band. The high level of concordance between the model and the protein is maintained when the ferrous carbonyl porphyrin thiolate is compared to ferrous carbonyl P-450$_{LM_2}$ (see Table 8, line 4, and Table 9, line 3). Again strong evidence is obtained for a thiolate ligand in P-450, since the N-methylimidazole ferrous carbonyl porphyrin (Table 9, line 4) gives a distinctly different spectrum. Indeed the latter complex, the thiol ferrous carbonyl porphyrin, and ferrous carbonyl P-420$_{LM_2}$ give virtually identical MCD spectra. Thus replacement of the thiolate in P-450 by imidazole, or even simple protonation of the thiolate is sufficient to destroy the unique "P-450" behavior and elicit the more normal "P-420" spectrum

4.7 Other Types of Spectroscopy

Additional spectroscopic methods which have been applied to studies of P-450 include nuclear magnetic resonance, infrared, and resonance Raman spectroscopy. The NMR studies have centered on paramagnetic iron-induced relaxation of protons on small molecules interacting with the P-450 active site. Thus Griffin and Peterson (116) determined the relaxation rates of bulk solution water protons in the presence of various complexes of ferric P-450$_{cam}$. They found that the calculated distance between the paramagnetic center and solvent protons was 4 to 6 Å in complexes with iron-coordinating ligands such as 4-phenylimidazole and about 5 to 9 Å in the substrate-bound pentacoordinate form. However in the absence of such ligands (i.e., the native, ferric low-spin form), the apparent distance to bulk water protons was only 2.0 to 2.6 Å. The interpretation was that a rapidly exchanging water molecule constitutes the sixth ligand in ferric, low-spin P-450$_{cam}$. In the presence of 4-phenylimidazole or in the substrate-bound form the water molecule is no longer directly coordinated to the iron. This conclusion is in good agreement with the results from studies of the electronic spectra of model compounds, which suggest that the sixth ligand is coordinated through an oxygen atom.

Similar NMR studies of bulk water proton relaxation have been performed on solubilized P-450$_{LM}$ (117). While no definite conclusion could be reached concerning possible axial ligation of a water molecule, the data favor a relatively open heme environment which could accommodate several water molecules. Another report of an NMR investigation of P-450$_{LM}$ by Novak et al. (118) showed, from measurements of relaxation times of protons on organic substrate molecules, that molecules such as aniline or imidazole probably bond directly to iron but that the sterically hindered base 2,6-xylidine is probably bound hydrophobically and can interact with the iron atom only by outer sphere mechanisms.

A limited amount of work involving vibrational spectroscopy has appeared. Infrared measurements were carried out on solubilized P-450$_{LM}$ (119). The carbonyl stretching frequency of ferrous carbonyl P-450$_{LM}$ was found to be 1949 cm^{-1}, and for the P-420 derivative the value was increased to 1966 cm^{-1}. Neither of these values were unusual with respect to the corresponding hemoglobin of myoglobin derivatives. Resonance Raman spectra have been reported only for P-450$_{cam}$. Ozaki et al. (120) found the so-called "oxidation-state marker" or Band IV at the unusually low frequency of 1346 cm^{-1} for ferrous P-450$_{cam}$. For other reduced hemeproteins this band is always found between 1355 and 1365 cm^{-1}. However the ferric substrate-free form absorbed at 1373 cm^{-1}, a frequency typical of other ferric hemeproteins. Champion and Gunsalus (121) determined that the resonant frequency for the corresponding band in substrate-bound P-450$_{cam}$ was 1368 cm^{-1}, which is anomalously low again. The significance of the unusual frequencies is not known, but thiolate ligation was suggested by both groups as a possible explanation.

To summarize the results of spectroscopic structural studies, the majority of evidence from several lines of research strongly suggests the presence of a thiolate ligand, probably from a deprotonated cysteine residue, as the fifth ligand to the heme iron in native P-450. In addition two independent approaches indicate an oxygen ligand, such as water or a protein hydroxyl or carboxyl group, as the sixth ligand in the native ferric, low-spin state. However a caveat is in order, inasmuch as no unequivocal, direct chemical evidence of the presence of a thiolate has yet been presented. Conceivably such final proof of structure may only be provided by X-ray crystallographic analysis.

5 MECHANISM OF OXYGEN ACTIVATION AND INSERTION INTO SUBSTRATES

5.1 Mechanistic Cycle

The current conceptual model of the process by which cytochrome P-450 catalyzes the reductive scission of oxygen and accompanying oxidation of carbon atoms is based on a number of years of investigations of P-450, other hemoproteins, and chemical model systems. This scheme, presented in Figure 12, takes into account the observed stoichiometry, the results of ^{18}O-labeling experiments, the regioselectivity of the hydroxylation, and partial loss of configuration during oxidation of prochiral centers. However not all steps and intermediates shown have been demonstrated, and the point indicated for entry of the protons is purely for convenience. The dashed line bisecting the circle is explained in Section 5.4.

For the present purpose we confine ourselves largely to a discussion of the reconstituted hydroxylation system comprised of purified P-450$_{LM_2}$ or P-450$_{LM_4}$, purified NADPH-cytochrome P-450 reductase,

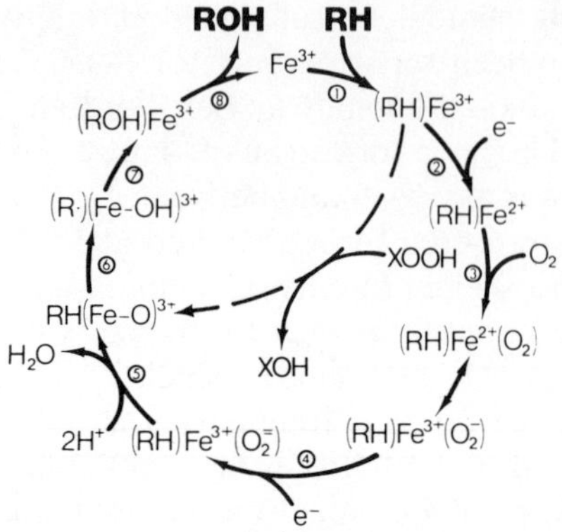

Figure 12. Catalytic cycle of P-450, showing intermediate reactions of the normal mechanism (solid arrows) and a tentative proposal for the hydroperoxide-dependent reactions (dashed arrow).

phospholipid, buffer, substrate, O_2, and NADPH (28). This reconstituted system is admittedly artificial; it differs principally in the ratio of P-450 to reductase, which in the microsomal membrane is about 20:1 (122) but is about 1:3 in the reconstituted system. The membranous milieu is another aspect of the microsomal system not present in the reconstituted system, and in addition other proteins may exert unappreciated effects in the microsomal membrane. Nonetheless four similarities foster confidence that the functioning of the reconstituted system is relevant to the physiological operation of P-450: (a) turnover numbers based on limiting reductase; (b) substrate specifically; (c) effects of inhibitors; and (d) efficiency of coupling of NADPH oxidation to product formation (16). In fact it seems certain that a detailed understanding of the microsomal processes will require fundamental knowledge of the chemistry of the individual proteins.

5.2 Individual Steps of the Cycle

Briefly, this model of the P-450 catalytic mechanism posits substrate binding to native ferric P-450 followed by reduction to the ferrous state, allowing oxygen binding. A second reduction induces splitting of the bound oxygen, one atom being extruded as water and the other being left behind as the "activated oxygen." The "activated oxygen" inserts into a proximate carbon–hydrogen bond forming the product alcohol, which is then released, thereby regenerating the original ferric state for a new catalytic cycle. The evidence or rationale for these steps will be presented below.

5.2.1 Binding of Substrate.

The preferred substrates of the $P\text{-}450_{LM}$ hydroxylation system are hydrophobic, apolar molecules. For instance, cyclohexane is one of the more rapidly hydroxylated substances. Polar substances such as alcohols have a much lower affinity for the active site, and ionic molecules may not be bound at all. Association of a hydrophobic molecule with $P\text{-}450_{LM}$ (step 1) causes a perturbation of the equilibrium between the high-spin and low-spin states which is reflected in a lowering of the ratio of absorbance at 418 nm to that at 394 nm. This alteration, which was first observed in microsomes, is referred to as a Type I spectral change when measured in the difference spectrum mode (123, 124) and has been used extensively as a convenient measure of the extent of substrate binding. The spin perturbation apparently arises from a displacement of the endogenous sixth ligand of the ferric low-spin form resulting in the high-spin pentacoordinate form. At least once inhibitor of the $P\text{-}450_{LM}$ monooxygenase, SKF-525A, is thought to attach tightly at the substrate-binding site, thereby competitively inhibiting the first step in the cycle.

5.2.2 First Reduction.

This reaction, step 2 in Figure 12, has been the subject of considerable study, both in microsomal and purified systems (15, 122, 125). The reduction is relatively easy to observe by rapid mixing of anaerobic, carbon monoxide-saturated solutions of the reductase–P-450 complex and of NADPH in a stopped-flow spectrophotometer. Since the binding of CO is extremely rapid, the observed change of absorbance at about 450 nm with time represents the entry of an electron into P-450 from NADPH via the reductase. Semilogarithmic plots of such data show the reaction to be first-order biphasic, but no rationale for this behavior has been communicated. Interestingly the rate constants for both phases are dependent on the presence of substrate with $P\text{-}450_{LM_2}$ and are large enough to account for the known turnover number only when substrate is present (125). Thus it appears that the first two steps of the reaction cycle are ordered in the sequence shown. Separate static oxidation–reduction titrations have shown that this step involves a reversible one-electron transfer (8). The midpoint potential for this couple with $P\text{-}450_{LM_2}$ is -326 mV but increases to -150 mV in the presence of CO (71). Such a low reduction potential may be a reflection of thiolate ligation.

5.2.3 Binding of Dioxygen.

Step 3 is particularly well established, since the ferrous dioxygen complex has been prepared and is stable enough to study in both the $P\text{-}450_{cam}$ (25, 88) and adrenal P-450 (126) systems (for the spectrum, see Table 2). In the absence of the appropriate reduced iron–sulfur redoxin, this complex decays by dissociation either of dioxygen or of superoxide ion, generating the ferrous or ferric protein, respectively. As expected, the dioxygen complex is diamagnetic and exhibits no EPR absorbance (88, 126). The reaction of dioxygen with ferrous P-450$_{LM_2}$ has also been investigated (127). The rapidity of the reactions necessitated the use of stopped-flow spectrophotometry. Two intermediates, presumably oxygenated, were observed. The first, complex I, was formed within the dead time of the instrument ($k_1 > 60,000$ min^{-1}), while the second, complex II, was generated by the first-order decay of complex I ($k_2 = 210$ min^{-1}). Neither complex has been identified, but the spectrum of complex I, with absorption maxima at 423, 560, and 680 (shoulder) nm, is not radically different from those of above-mentioned ferrous dioxygen complexes. Complex II disappeared in another first-order process ($k_3 = 12$ min^{-1}) to generate the ferric cytochrome. Since complex I does not dissociate into superoxide and ferric P-450, we may postulate that disproportionation occurs between two or more molecules of the reduced oxygenated complex to generate ferric P-450 and a two-electron reduced oxycomplex. The latter complex could, through additional steps, oxidize substrate, release hydrogen peroxide, or further disproportionate (yielding water) to ultimately form the ferric cytochrome as the final product. At some stage in this process the mixture of an oxygenated or higher oxidation state heme and the ferric P-450 could give rise to the observed spectrum of complex II. No irreversible oxidation of the hemoprotein appears to occur. The postulated disproportionation is not unreasonable within a large aggregate of P-450$_{LM_2}$ molecules. Experiments similar to these were carried out at low temperatures with microsomal suspensions (128). Two "oxy-ferro" complexes were observed after mixing O$_2$ with reduced microsomes at $-45°C$, but the two intermediates do not appear to be identical to those seen using P-450$_{LM_2}$.

5.2.4 Second Reduction. Introduction of a reducing equivalent into the ferrous dioxygen complex, step 4, has not been measured or even observed with P-450$_{LM_2}$. Attempts to discern this reduction step with ferrous dioxygen P-450$_{cam}$ have resulted only in the observation of the rapid collapse of that complex with simultaneous appearance of the ferric form of the enzyme, no intermediates being detected (129, 130). We are obligated to include this step at this point since two reducing equivalents are required to satisfy the reaction stoichiometry, and the second may only enter the scheme after dioxygen binding. Estabrook and co-workers (75, 131–133) have suggested that in microsomes the second electron may derive from NADH via the cytochrome b_5 electron transport chain. This proposal is based on the observation that the rate of substrate oxidation when both NADH and NADPH are present is greater than the sum of the rates for NADH- or NADPH-dependent oxidations determined separately. In this "synergistic" scheme, cytochrome b_5 is thought to provide electrons to the ferrous dioxygen complex faster than the NADPH-cytochrome P-450 reductase is able to, thereby increasing the rate of turnover of P-450. Implicit in this proposition is the assumption that step 4 is at least partially rate limiting in the catalytic cycle. Subsequent work by Imai and Sato (134) demonstrated a similar "synergism" in a reconstituted system containing P-450$_{LM_2}$, the P-450 reductase, and in some cases cytochrome b_5 and its associated flavoprotein reductase. In addition to rate enhancement associated with the presence of NADH, they observed a marked improvement in the degree of coupling of NAD(P)H oxidation to substrate oxidation. The latter effect was attributed to an effector role of the cytochrome b_5 independent of its postulated electron transfer role.

Several other groups (76, 135, 136), working with microsomes, have suggested an alternative explanation of NADH–NADPH "synergism." According to this hypothesis, P-450 and cytochrome b_5 compete for electrons from NADPH-cytochrome P-450 reductase. Since the reductase is the rate-limiting component of the hydroxylase system, shunting of electrons through cytochromes b_5 effectively retards the overall P-450-dependent hydroxylation rate. The reduced cytochrome b_5 can recycle through autoxidation or reduction of the desaturase (137). When NADH is added to microsomes or when NADH, cytochrome b_5 reductase, and cytochrome b_5 are added to a reconstituted system, the much faster reduction of cytochrome b_5 at the expense of NADH causes it to be primarily in the ferrous form in the steady state. In this circumstance no reducing equivalents are shunted away from P-450, and the hydroxylation system appears to cycle more rapidly.

The proposed structure for the two-electron reduced P-450-oxygen complex is a ferric iron-coordinated peroxide complex of unspecified protonation state. The intermediacy of this species is only speculative, but its inclusion in the diagram provides a convenient point for the postulated entry of certain peroxides into the cycle (see Section 5.4). The generation of hydrogen peroxide by microsomes and by the reconstituted system (16) in the presence of NADPH and oxygen may result from dissociation of this complex, although the dissociation and subsequent disproportionation of superoxide ion from the ferrous dioxygen complex is probably more likely (138).

5.2.5 Splitting of the Oxygen–Oxygen Bond. Step 5, the generation of an "active oxygen" by heterolytic cleavage of the peroxide bond, is the most speculative and least understood process in the entire scheme. The overall result of the P-450-catalyzed reaction with a hydrocarbon substrate is the insertion of an oxygen atom into a carbon–hydrogen bond on the substrate, eq. (1). The formal equivalence of this process to the insertion of carbenes and nitrenes into carbon–hydrogen bonds prompted Hamilton (139) to use the term "oxenoid mechanism" in regard to monooxygenases in general and to P-450 in particular. Similarly Ullrich (140) has referred to P-450 as an "oxene transferase," both terms reflecting the ability of the enzyme to remove an oxygen atom from molecular oxygen and insert it into the substrate through the intermediacy of a covalently bound enzyme-oxygen atom species. At no time is a free oxygen atom thought to be present. This process is called the "activation of oxygen" because ground-state molecular oxygen itself is unreactive toward most organic substances (139).

In addition to the conversion of alkanes to alcohols, P-450 performs other oxidations formally identical to those typical of oxenoid reagents. For instance, alkenes are converted to epoxides and alkyl sulfides are oxidized to sulfoxides (34). Furthermore certain aromatic substrates exhibit a migration of substituents which has been termed the "NIH shift," and this reaction has been shown to be a consequence of the intermediacy of an arene oxide (141, 142). Possible structures for a P-450 oxenoid intermediate capable of performing the reactions listed above include a peracid, a peramide, and an iron–oxygen atom complex (139).

We have chosen to represent the "activated oxygen" as an oxygen atom coordinated to ferric iron since this species would be most expected to exhibit the reaction characteristics observed with P-450, as detailed later. The overall charge on this two-atom unit, neglecting the contributions of the porphyrin and the thiolate, is therefore $3+$. Thus the designation $(Fe-O)^{3+}$ represents a ferric iron-coordinated oxygen atom of unspecified local electron density. We might as easily write $(Fe^{IV}-O^-)^{3+}$ or $(Fe^V-O^{2-})^{3+}$, using the dash to signify a coordinate-covalent bond.

The conversion of the iron–peroxide complex to the oxenoid species involves protonation and heterolytic cleavage of the oxygen–oxygen bond with extrusion of water such that the remaining oxygen atom contains the two oxidation equivalents previously associated with the peroxide. This process has been likened to that by which hydrogen peroxide reacts with peroxidase to form Compound I, which is two oxidation equivalents above the native, ferric state (143). The analogy is strengthened by the fact that Compound I apparently contains only one oxygen atom from the original peroxide (144–146). A role for the thiolate ligand in the splitting of the peroxide bond in the P-450 reaction has been proposed in which the peroxide bond is weakened by charge repulsion in the neighborhood of the electron-rich iron (115). However it may be noted that thiolate is not a required ligand in peroxidase.

5.2.6 Hydrogen Abstraction and Radical Recombination. One would expect the iron-oxenoid species to be electrophilic and very reactive toward oxidizable organic molecules. The oxenoid intermediate of P-450 appears to be special in this respect, because P-450 is the only hemoprotein capable of hydroxylating an alkane at an unactivated carbon–hydrogen bond. Even peroxidase Compound I is unable to achieve this reaction, despite the fact that similar oxidized intermediates have been proposed for both P-450 and peroxidase. It is intriguing of course to speculate on a unique role of the proposed thiolate ligand in modulating the reactivity of the oxidant, but no convincing arguments have been offered to substantiate such a role. A number of chemical model systems have been developed over the past decade which are able to oxidize various organic compounds, usually employing hydrogen peroxide or molecular oxygen as the oxidant and source of oxygen atoms. Particularly relevant to the problem of the P-450-catalyzed hydroxylations are those based on metal complexes, of which there are two types. Representative of the first type are those which utilize molecular oxygen and a reductant, either a low-valence metal or an organic reductant (147–151). Those of the second type are modified Fenton systems, one of which has been particularly well characterized mechanistically by Groves and coworkers (152, 153). The system consists of ferrous perchlorate, hydrogen peroxide, and cyclohexanol as the hydroxylatable substrate, in acetonitrile as solvent. Cyclohexanol was hydroxylated regio- and stereoselectively, the cis-1,3-diol accounting for more than 70% of the product (153). In addition, ^{18}O-labeling experiments established that the inserted oxygen atom originated from H_2O_2, and competitive oxidation of cyclohexanol and 2,2,6,6-tetradeuterocyclohexanol demonstrated the existence of a substantial isotope effect (k_H/k_D = 3.2). Furthermore it appeared that cis-hydrogen removal resulted in cis-diol formation while trans-hydrogen removal afforded trans-diol. These experimental results were accommodated by a mechanism involving a cyclohexanol-coordinated iron-bound oxygen atom as the key intermediate in a two-step hydroxylation process. In the first step hydrogen abstraction by the iron-oxenoid species led to a transient carbon radical juxtaposed to an iron-bound hydroxyl radical. In the second step either direct combination of the radicals or simultaneous one-electron oxidation of the carbon radical and collapse of the hydroxide–carbenium ion pair generated an alcohol functional group. This two-step hydrogen abstraction–radical recombination process was termed "oxygen rebound" (152).

Aliphatic hydroxylation by P-450 may also proceed by "oxygen rebound." Thus in Figure 12, steps 6 and 7 represent the hydrogen abstraction and radical recombination steps, respectively. Although the iron-oxenoid species depicted may not actually be the "activated oxygen," the true intermediate appears to react in this two-step fashion. The carbon–hydrogen bond on the substrate must be broken in one of four ways: (a) as a radical pair (homolytically); (b) as a carbanion and proton (heterolytically); (c) as a carbenium ion and hydride ion (heterolytically); or (d) by a direct insertion of oxygen (concertedly). Hydroxylation of small aliphatic molecules by P-450 results in a preference for 3° positions, with 2° positions being less reactive, and with 1° hydrogens disfavored (154). Such regioselectivity is characteristic of hydrogen abstractions by moderately active, selective radicals but would not be expected with carbanions or with direct insertions.

Deuterium isotope effects provide additional support for an initial hydrogen abstraction. When deuterated and undeuterated molecules are compared with respect to overall rates of hydroxylation, little or no kinetic isotope effect is seen (155–157) inasmuch as hydrogen abstraction is not rate limiting. However when a substrate molecule contains both hydrogen and deuterium in hydroxylatable positions, large isotope effects may be deduced (158). Such an *intramolecular* isotope effect was invoked for the P-450-catalyzed hydroxylation of *exo,exo,exo,exo*-tetradeuteronorbornane to yield a mixture of *endo*- and *exo*-2-norborneol (157). The *exo/endo* ratio was 3.4:1 with norbornane but dropped to 0.76:1 with the deuterated analog, indicative of a large preference for hydrogen over deuterium abstraction ($k_H/k_D = 11$). Large isotope effects are frequently observed with selective hydrogen abstractors (159) but are not expected with direct insertions into carbon–hydrogen bonds. Mass-spectral analysis of the alcohols from the P-450-catalyzed hydroxylation of tetradeuteronorbornane showed the presence of small amounts of *endo*-deuterium in the *exo*-alcohol and a deficit of similar magnitude of *exo*-deuterium in the *endo*-alcohol. Clearly the reaction did not proceed with 100% retention of configuration, but rather when the deuterium was abstracted from the *exo* face the hydroxyl radical occasionally recombined on the opposite side. This result strongly suggests the presence of a discrete carbon radical which exists long enough to pull partly away from the enzyme-bound oxidant and sometimes present the other face for recombination. Furthermore the crossover from *exo* to *endo* rules out the intermediacy of a 2-norbornyl carbenium ion, since such cations are never captured in the *endo* position (160, 161). The "oxygen rebound" mechanism with partial epimerization of the oxidized carbon center is depicted in Figure 13.

The picture which emerges from the indirect study of the P-450 oxenoid intermediate by product analysis is that the "activated oxygen" is a moderately active, electrophilic, hydrogen-abstracting species which hydrox-

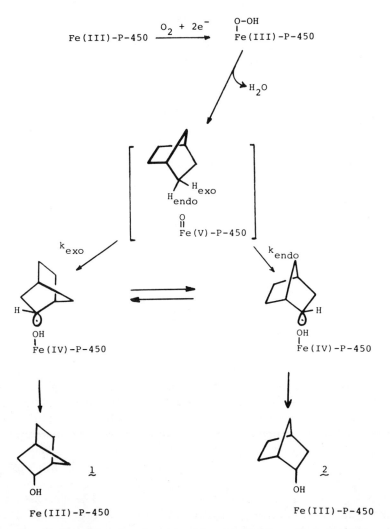

Figure 13. Hydroxylation of norbornane catalyzed by P-450$_{LM2}$, illustrating "oxygen rebound" and partial epimerization, taken from reference 157.

ylates alkanes in a two-step process. This description fits rather well the reported behavior of triplet oxygen atoms from γ-radiolysis of carbon dioxide solutions (162). These oxygen atoms were able to hydroxylate alkanes with a regioselectivity similar to that of P-450. Interestingly these hydroxylations occurred with a 7 to 17% loss of configuration at the oxidized carbon, which is similar to that seen with P-450. An "oxygen rebound" mechanism was proffered to rationalize the results. Thus study of the P-450 "activated oxygen" by product analysis suggests a hydrogen abstractor acting in a two-step mechanism, and the behavior of model systems makes reasonable its designation as a ferric iron-bound oxygen atom.

5.2.7 Dissociation of Product. The last step in the cycle is the dissociation of the product alcohol from the P-450 active site (Fig. 12, step 8). This process apparently occurs before a new reaction cycle begins, since diols are not observed in alkane hydroxylations. The hydroxyl group will certainly be coordinated to the iron immediately after its formation; this may have two effects which inhibit a second hydroxylation. First, the low-spin hexacoordinate iron–alcohol complex may be sluggishly reduced; and second, dioxygen binding to the reduced iron cannot occur until the sixth ligation position is vacated. Thermodynamically the polar alcohol will probably be bound with less affinity than its parent alkane, so that replacement of the oxidized molecule with a fresh substrate molecule is favored. Of course in most instances, once the alcohol dissociates from the enzyme surface, the much greater concentration of unaltered substrate eliminates any possibility of a second hydroxylation.

5.3 Rate-Limiting Step

Discussion of the rate-limiting step in the liver microsomal P-450 system is complicated by the fact that, while many different hydrophobic compounds can be oxidized by P-450, the rates at which the oxidations are accomplished vary enormously, with no relationship discerned to date between substrate structure and V_{max}. However with the more rapidly hydroxylated compounds, only three of the eight steps discussed in Section 5.2 can reasonably be considered as candidates for the rate-limiting step. The derivation of this statement follows.

Step 1, binding of substrate, is easily dismissed. The association of benzphetamine with P-450$_{LM}$ proceeds with a second-order rate constant of 9.3×10^4 M^{-1} sec^{-1} (163), and the corresponding constant for the binding of camphor to P-450$_{cam}$ is 4.1×10^6 M^{-1} (164). Step 2, the reductase-mediated reduction of the ferric heme, is relatively fast as well, as judged by the rate of appearance of the ferrous carbonyl cytochrome when NADPH is mixed with a reconstituted cytochrome P-450– reductase–phospholipid system anaerobically in the presence of carbon monoxide. The first-order rate constants determined for this process range from about 100 to 400 min^{-1}, depending on the system studied, but the corresponding turnover numbers for the complete hydroxylation systems are in the range of 22 to 70 min^{-1} (15, 125). Thus while this reaction is complicated by biphasic, first-order kinetics, and while the rate constants are not orders of magnitude greater than the turnover number of the hydroxylase, nonetheless this process cannot be considered rate limiting.

The association of dioxygen with the cytochrome (step 3) is clearly not rate limiting, the second-order rate constant having been estimated as $7.7 \times 10^5 \ M^{-1} \ sec^{-1}$ in the case of P-450$_{cam}$ (88). The hydrogen abstraction (step 6) may be safely ruled out since there is no correlation of turnover numbers with ease of hydrogen abstraction in hydrocarbons, and there is no significant kinetic isotope effect when deuterated hydrocarbons are compared to ordinary hydrocarbons (155, 157). Step 7, the radical recombination reaction, is expected to be controlled by the rates of collision of the proximate radicals, such reactions generally being considered to involve little or no activation energy (165). The partial epimerization of 2-norbornyl radicals during hydroxylation (36) may be evidence that the rates of recombination and diffusional separation are comparable. Clearly the collapse of the radical pair must be extremely rapid.

Of the remaining three steps, the second reduction, the generation of the "active oxygen," and the dissociation of the product, little can be said. Any of them could plausibly be slow enough to account for turnover, or perhaps a combination of similar rate constants leading to an apparent rate constant under steady-state conditions must be invoked. Nothing is known about the nature of the second reduction with the microsomal enzyme system. In the bacterial P-450$_{cam}$ hydroxylase system, this rate agrees well with the observed turnover number for camphor 5-*exo*-hydroxylation and appears to be the rate-determining step on that basis (129, 130). However, extrapolation of this conclusion to the microsomal enzyme is not warranted presently, since the microsomal system shows less than 10% of the P-450$_{cam}$ turnover rate.

Concerning the other two potential rate-limiting steps, again no conclusions may be drawn. While the intermediacy of a discrete "activated oxygen" may be inferred from regioselectivity data (154) and the observation of internal hydrogen isotope selectivity (157), no direct physical evidence of its existence is yet available. Therefore any conclusions concerning the rate of formation of the "activated oxygen" will depend on knowledge of the rates of all other processes in the mechanistic cycle.

The rate of dissociation of product, on the other hand, is in principle amenable to direct determination by measurements of the association rate of product with native enzyme (k_{on}) and the dissociation constant of the enzyme-product complex. By an "order of magnitude" calculation, assuming $k_{on} = 10^5 \, M^{-1} \, sec^{-1}$ and $K_d = 10^{-6} \, M$, we have shown that the dissociation rate (k_{off}) may be small enough to limit the overall rate of P-450 cycling. Postulation of product release as rate limiting offers a partial explanation of the observed variation in V_{max} with different substrates, since k_{off} should be a function of product structure. Therefore product dissociation rates should be measured in order to evaluate this postulate.

The turnover number of the reductase–P-450 complex is the same in the microsomal membrane and in the reconstituted system. In liver microsomes from phenobarbital-induced rabbits, the initial rates of NADPH oxidation and of benzphetamine demethylation, respectively, are 9 and 4 nmole/min/nmole of P-450 at 30°C with saturating concentrations of these substances (16). However, since there is a limiting amount of reductase in such microsomes (P-450:reductase ratio is 20:1), these rates must be multiplied by 20 in order to express them on the basis of the actual limiting component. Thus the rates are 180 nmole/min/nmole of reductase for NADPH oxidation and 80 nmole/min/nmole reductase for formaldehyde accumulation, the overall efficiency of coupling being 0.44 mole CH_2O formed per mole NADPH consumed. For comparison, Vermilion and Coon (64) measured rates of NADPH oxidation and benzphetamine demethylation under similar conditions using the reconstituted system with the purified reductase rate limiting. In these experiments NADPH oxidation proceeded at 197 nmole/min/nmole reductase, and formaldehyde accumulated at the rate of 92 nmole/min/nmole reductase, giving an efficiency of coupling of 0.47 mole CH_2O per mole NADPH oxidized. These results suggest that the rate-limiting step in the P-450 catalytic cycle is the same in both the microsomal and reconstituted systems.

5.4 Hydroperoxide-Supported Hydroxylations

As indicated in Figure 12, it is possible to bypass the normal reduction and dioxygen-binding steps by adding a peroxide (symbolized as XOOH) to the substrate-bound P-450. However it is not known if any common intermediates exist. This shunt was first observed in microsomes by Hrycay and O'Brien (166), who measured rates of tetramethyl-p-phenylenediamine oxidation by organic hydroperoxides. An earlier paper by Chen and Lin (167) reported that tetralin hydroperoxide was an intermediate in the hydroxylation of tetralin, but the significance of this finding is not clear. A large number of publications following these initial observations established that the microsomal peroxidase was able to perform the same substrate hydroxylations catalyzed by P-450 and was probably identical with the latter enzyme (19, 20, 168–171). Additional work pointed out that these oxidation reactions could also be observed with other types of oxidants, including sodium periodate (169, 170), sodium chlorite (169, 172), and iodosobenzene (140, 172).

Some related work has appeared based on purified forms of P-450, including P-450$_{cam}$ (173) and adrenal P-450 (126). The only investigation utilizing purified P-450$_{LM}$ has been that of Nordblom, White, and Coon (23). It was found that NADPH–cytochrome P-450 reductase and NADPH could be replaced by certain peroxides in the reconstituted system, although the requirement for phospholipid remained. Several oxidants, including alkyl hydroperoxides, hydrogen peroxide, peroxy acids, and sodium chlorite, were shown to be capable of demethylating benzphetamine. Rather high concentrations of some of these agents were required to effect the reaction, and consequently irreversible oxidation of the heme always accompanied the desired reaction. However it was possible to minimize destruction in some instances, so that steady-state kinetics of the reaction could be studied, the accumulation of product with time being linear for at least 2 min. Thus it was shown that the reaction depended on the presence of P-450 since P-420 or other ferric complexes were ineffective. Typical P-450 inhibitors showed the expected decrease of reaction rates in the peroxide system, but there was no rate decrease when the system was made anaerobic, either in the presence or absence of carbon monoxide. Clearly molecular oxygen is not a participant in the reaction, and the ferrous oxidation state does not appear to occur during the catalytic cycle.

The stoichiometry of the cumene hydroperoxide-promoted demethylation of N,N-dimethylaniline was shown to be 1 mole N-methylaniline and 1 mole formaldehyde produced for each mole cumene hydroperoxide creduced to the corresponding alcohol:

$$PhN(CH_3)_2 + PhC(CH_3)_2OOH$$
$$\rightarrow PhNHCH_3 + HCHO + PhC(CH_3)_2OH \quad (8)$$

Organic hydroperoxides also can hydroxylate alkanes with P-450$_{LM_2}$ present, although this reaction has not been detected when H_2O_2 is the oxidant. When cyclohexane was hydroxylated at the expense of cumene hydroperoxide in a buffer containing $H_2^{18}O$, the resulting cyclohexanol was found by mass spectroscopy to have derived less than 10% of its oxygen atoms from water (23). This result, and those preceding it, suggest the intermediacy of an enzyme-bound oxenoid species able to transfer one oxygen atom from the peroxide to the substrate. Rahimtula et al. (76) reported a dramatic spectral change upon mixing microsomes and cumene hydroperoxide and suggested that this may reflect the presence of a higher oxidation state oxygen complex of the microsomal P-450. A similar change is observed when P-450$_{LM_2}$ is combined with alkyl hydroperoxides or peroxy acids (but not with hydrogen peroxide). As shown in Figure 14, the difference spectrum compared to native P-450$_{LM_2}$ shows a trough at about 415 nm and a peak near 436 nm. Although the rapidly formed intermediate begins to decay within a few seconds to degraded products, the use of stopped-flow spectrophotometry makes it possible to reconstruct the absolute spectrum, as shown in Figure 15. The Soret band is red-shifted to about 425 nm and exhibits a decreased absorbance. This spectrum is not at all like that of horseradish peroxidase, Compound I, which has a Soret band at 400 nm (144) but shows vague similarity to the spectrum of cytochrome c–peroxidase complex II, with a Soret band at 418 nm (175). No relation between the P-450$_{LM_2}$ peroxide-induced spectrum and the hydroxylation reaction has been demonstrated, and the fact that the intermediate cannot be formed stoichiometrically from added peroxide

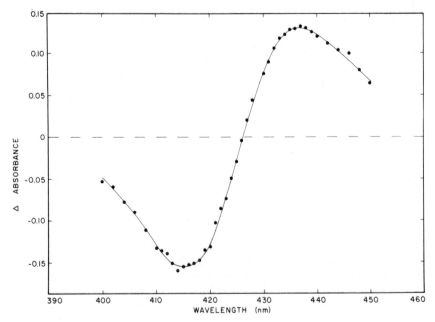

Figure 14. Difference spectrum relative to native P-450$_{LM_2}$ resulting from reaction of P-450$_{LM_2}$ (2.7 μM) with m-chloroperoxybenzoic acid (20 μM) in phosphate buffer. (0.05 M, pH 7.0) containing dilauroyl-GPC (50 μg per ml) at 10°, taken from reference 174.

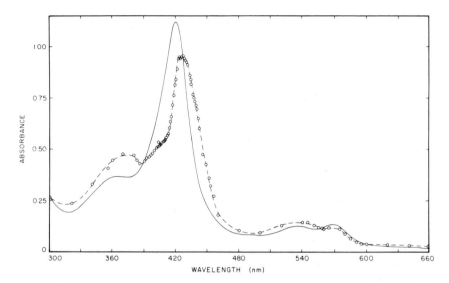

Figure 15. Absolute spectrum of intermediate formed from reaction of P-450$_{LM2}$ with *m*-chloroperoxybenzoic acid, reconstructed from data in Figure 14. Solid line: native P-450$_{LM2}$. Circles: intermediate. Taken from reference 174.

(174), as can Compound I, makes it unlikely that this spectrum represents an iron-oxenoid compound. We are now investigating the nature of this compound as well as the larger question of the relationship of the mechanism of the peroxide-supported hydroxylation to that of the NADPH- and O$_2$-supported reaction.

ACKNOWLEDGMENT

Preparation of this review was aided by Grant AM-10339 from the Institute for Arthritis and Metabolic Diseases, United States Public Health Service, and by Grant PCM76-14947 from the National Science Foundation.

REFERENCES

1. K. J. Ryan and L. L. Engel, *J. Biol. Chem.*, **225**, 103 (1957).
2. M. Klingenberg, *Arch. Biochem. Biophys.*, **75**, 376 (1958).
3. D. Garfinkel, *Arch. Biochem. Biophys.*, **77**, 493 (1958).
4. M. Katagiri, B. N. Ganguli, and I. C. Gunsalus, *J. Biol. Chem.*, **243**, 3543 (1968).
5. R. W. Estabrook, D. Y. Cooper, and O. Rosenthal, *Biochem. Z.*, **338**, 741 (1963).

6. T. Omura, R. Sato, D. Y. Cooper, O. Rosenthal, and R. W. Estabrook, *Fed. Proc.*, **24**, 1181 (1965).
7. T. Omura and R. Sato, *J. Biol. Chem.*, **239**, 2370 (1964).
8. J. A. Peterson, R. E. White, Y. Yasukochi, M. L. Coomes, D. H. O'Keeffe, R. E. Ebel, B. S. S. Masters, D. P. Ballou, and M. J. Coon, *J. Biol. Chem.*, **252**, 4431 (1977).
9. J. R. Gillette, *Adv. Pharmacol.*, **4**, 219 (1966).
10. R. Kato, K. Iwasaki, T. Shiraga, and H. Noguchi, *Biochem. Biophys. Res. Commun.*, **70**, 681 (1976).
11. R. Kato, K. Iwasaki, and H. Noguchi, *Biiochem. Biophys. Res. Commun.*, **72**, 267 (1976).
12. A. Y. H. Lu and M. J. Coon, *J. Biol. Chem.*, **243**, 1331 (1968).
13. M. J. Coon and A. Y. H. Lu, in J. R. Gillette, A. H. Conney, G. J. Cosmides, R. W. Estabrook, J. R. Fouts, and G. J. Mannering, Eds., *Microsomes and Drug Oxidations*, Academic Press, New York, 1969, p. 151.
14. A. Y. H. Lu, K. W. Junk, and M. J. Coon, *J. Biol. Chem.*, **244**, 3714 (1969).
15. H. W. Strobel, A. Y. H. Lu, J. Heidema, and M. J. Coon, *J. Biol. Chem.*, **245**, 4851 (1970).
16. G. D. Nordblom and M. J. Coon, *Arch. Biochem. Biophys.*, **180**, 343 (1977).
17. F. F. Kadlubar, K. C. Morton, and D. M. Ziegler, *Biochem. Biophys. Res. Commun.*, **54**, 1255 (1973).
18. A. D. Rahimtula and P. J. O'Brien, *Biochem. Biophys. Res. Commun.*, **60**, 440 (1974).
19. Å. Ellin and S. Orrenius, *FEBS Lett.*, **50**, 378 (1975).
20. A. D. Rahimtula and P. J. O'Brien, *Biochem. Biophys. Res. Commun.*, **62**, 268 (1975).
21. E. G. Hrycay, J.-Å. Gustafsson, M. Ingelman-Sundberg, and L. Ernster, *FEBS Lett.*, **56**, 161 (1975).
22. A. A. Akhrem, D. I. Metelitza, S. M. Bielski, P. A. Kiselev, M. E. Skurko, and S. A. Usanov, *Croat. Chem. Acta*, **49**, 223 (1977).
23. G. D. Nordblom, R. E. White, and M. J. Coon, *Arch. Biochem. Biophys.*, **175**, 524 (1976).
24. C. Yu and I. C. Gunsalus, *Biochem. Biophys. Res. Commun.*, **40**, 1431 (1970).
25. I. C. Gunsalus, C. A. Tyson, and J. D. Lipscomb, in *Oxidases and Related Redox Systems, Vol. 2*, T. E. King, H. S. Mason, and M. Morrison, Eds., University Park Press, Baltimore, 1973, p. 583.
26. I. C. Gunsalus and G. C. Wagner, *Methods Enzymol.*, **52**, 166 (1978).
27. D. H. O'Keeffe, R. E. Ebel, and J. A. Peterson, *Methods Enzymol.*, **52**, 151 (1978).
28. D. A. Haugen, T. A. van der Hoeven, and M. J. Coon, *J. Biol. Chem.*, **250**, 3567 (1975).
29. T. A. van der Hoeven and M. J. Coon, *J. Biol. Chem.*, **249**, 6302 (1974).
30. T. A. van der Hoeven, D. A. Haugen, and M. J. Coon, *Biochem. Biophys. Res. Commun.*, **60**, 569 (1974).
31. D. A. Haugen and M. J. Coon, *J. Biol. Chem.*, **251**, 7929 (1976).
32. M. J. Coon, D. P. Ballou, D. A. Haugen, S. O. Krezoski, G. D. Nordblom, and R. E. White, in *Microsomes and Drug Oxidations*, V. Ullrich, I. Roots, A. G. Hildebrandt, R. W. Estabrook, and A. H. Conney, Eds., Pergamon Press, Oxford, 1977, p. 131.
33. *Enzyme Nomenclature*, American Elsevier, New York, 1972, p. 24.

REFERENCES

34. M. J. Coon, J. L. Vermilion, K. P. Vatsis, J. S. French, W. L. Dean, and D. A. Haugen, in *Drug Metabolism Concepts*, D. M. Jerina, Ed., American Chemical Society Symposium Series, No. 44, 1977, p. 46.
35. D. Ryan, A. Y. H. Lu, J. Kawalek, S. B. West, and W. Levin, *Biochem. Biophys. Res. Commun.*, **64**, 1134 (1975).
36. J. C. Kawalek, W. Levin, D. Ryan, P. E. Thomas, and A. Y. H. Lu, *Mol. Pharmacol.*, **11**, 874 (1975).
37. Y. Imai and R. Sato, *Biochem. Biophys. Res. Commun.*, **60**, 8 (1974).
38. C. Hashimoto and Y. Imai, *Biochem. Biophys. Res. Commun.*, **68**, 821 (1976).
39. R. L. Norman, E. F. Johnson, and U. Muller-Eberhard, *J. Biol. Chem.*, **253**, 8640 (1978).
40. E. Arinc and R. M. Philpot, *J. Biol. Chem.*, **251**, 3213 (1976).
41. F. P. Guengerich, *Mol. Pharmacol.*, **13**, 911 (1977).
42. A. Y. H. Lu and S. B. West, *Pharmacology and Therapeutics* A, Vol. 2, Pergamon Press, London, 1978, p. 337.
43. W. L. Dean and M. J. Coon, *J. Biol. Chem.*, **252**, 3255 (1977).
44. K. Dus, W. J. Litchfield, A. G. Miguel, T. A. van der Hoeven, D. A. Haugen, W. L. Dean, and M. J. Coon, *Biochem. Biophys. Res. Commun.*, **60**, 15 (1974).
45. M. Tanaka, S. Zeitlin, K. T. Yasunobu, and I. C. Gunsalus, in *Iron and Copper Proteins*, K. T. Yasunobu, H. F. Mower, and O. Hayaishi, Eds., Plenum Press, New York, p. 263.
46. K. Dus, M. Katagiri, C-A. Yu, D. L. Erbes, and I. C. Gunsalus, *Biochem. Biophys. Res. Commun.*, **40**, 1423 (1970).
47. D. A. Haugen, L. G. Armes, K. T. Yasunobu, and M. J. Coon, *Biochem. Biophys. Res. Commun.*, **77**, 967 (1977).
48. G. Blobel and B. Dobberstein, *J. Cell. Biol.*, **67**, 835 (1975).
49. B. W. Griffin, J. A. Peterson, J. Werringloer, and R. W. Estabrook, *Ann. N.Y. Acad. Sci.*, **244**, 107 (1975).
50. H. S. Mason, J. C. North, and M. Vanneste, *Fed. Proc.*, **24**, 1172 (1965).
51. C. R. E. Jefcoate and J. L. Gaylor, *Biochemistry*, **8**, 3464 (1969).
52. J. Peisach and W. E. Blumberg, *Proc. Natl. Acad. Sci. USA*, **67**, 172 (1970).
53. D. W. Nebert, J. Robinson, and H. Kon, *J. Biol. Chem.*, **248**, 7637 (1973).
54. R. Tsai, C-A. Yu, I. C. Gunsalus, J. Peisach, W. Blumberg, W. H. Orme-Johnson, and H. Beinert, *Proc. Natl. Acad. Sci. USA*, **66**, 1157 (1970).
55. J. L. Vermilion and M. J. Coon, *Biochem. Biophys. Res. Commun.*, **60**, 1315 (1974).
56. B. L. Horecker, *J. Biol. Chem.*, **183**, 593 (1950).
57. C. H. Williams, Jr., and H. Kamin, *J. Biol. Chem.*, **237**, 587 (1962).
58. A. H. Phillips and R. G. Langdon, *J. Biol. Chem.*, **237**, 2652 (1962).
59. M. J. Coon, H. W. Strobel, and R. F. Boyer, *Drug Metab. Disp.*, **1**, 92 (1973).
60. M. J. Coon, D. A. Haugen, F. P. Guengerich, J. L. Vermilion, and W. L. Dean, in *The Structural Basis of Membrane Function*, Y. Hatefi and L. Djavadi-Ohaniance, Eds., Academic Press, New York, 1976, p. 409.
61. T. Iyanagi and H. S. Mason, *Biochemistry*, **12**, 2297 (1973).
62. Y. Yasukochi and B. S. S. Masters, *J. Biol. Chem.*, **251**, 5337 (1976).

63. J. D. Dignam and H. W. Strobel, *Biochemistry*, **16**, 1116 (1977).
64. J. L. Vermilion and M. J. Coon, *J. Biol. Chem.*, **253**, 2694 (1978).
65. J. S. French and M. J. Coon, *Arch. Biochem. Biophys.*, **195**, 565 (1979).
66. B. S. S. Masters, H. Kamin, Q. H. Gibson, and C. H. Williams, Jr., *J. Biol. Chem.*, **240**, 921 (1965).
67. B. S. S. Masters, M. H. Bilimoria, H. Kamin, and Q. H. Gibson, *J. Biol. Chem.*, **240**, 4081 (1965).
68. B. S. S. Masters, R. A. Prough, and H. Kamin, *Biochemistry*, **14**, 607 (1975).
69. T. Iyanagi, N. Makino, and H. S. Mason, *Biochemistry*, **13**, 1701 (1974).
70. J. L. Vermilion and M. J. Coon, *J. Biol. Chem.*, **253**, 8812 (1978).
71. F. P. Guengerich, D. P. Ballou, and M. J. Coon, *J. Biol. Chem.*, **250**, 7405 (1975).
72. A. P. Autor, R. M. Kaschnitz, J. K. Heidema, and M. J. Coon, *Mol. Pharmacol.*, **9**, 93 (1973).
73. Y. L. Chiang and M. J. Coon, *Arch. Biochem. Biophys.*, **195**, 178 (1979).
74. A. H. Conney, R. R. Brown, J. A. Miller, and E. C. Miller, *Cancer Res.*, **17**, 628 (1957).
75. A. Hildebrandt and R. W. Estabrook, *Arch. Biochem. Biophys.*, **143**, 66 (1971).
76. H. Staudt, F. Lichtenberger, and V. Ullrich, *Eur. J. Biochem.*, **46**, 99, (1974).
77. I. Jansson and J. B. Schenkman, *Mol. Pharmacol.*, **9**, 840 (1973).
78. A. Y. H. Lu, W. Levin, S. B. West, M. Vore, D. Ryan, R. Kuntzman, and A. H. Conney, in *Cytochromes P-450 and b_5: Structure, Function, and Interaction*, D.Y. Copper, O. Rosenthal, R. Snyder, and C. Witmer, Eds., Plenum Press, New York, 1975, p. 447.
79. J. P. Hummel and W. J. Dreyer, *Biochim. Biophys. Acta*, **63**, 530 (1962).
80. P. P. M. Bonsen, G. H. de Haas, W. A. Pieterson, and L. L. M. van Deenen, *Biochim. Biophys. Acta*, **270**, 364 (1972).
81. T. C. Pederson, J. A. Buege, and S. D. Aust, *J. Biol. Chem.*, **248**, 7134 (1973).
82. B. S. S. Masters, C. H. Williams, Jr., and H. Kamin, *Methods Enzymol.*, **10**, 565 (1967).
83. J. S. French, Doctoral Thesis, University of Michigan, 1979.
84. M. J. Coon, *Nutrition Rev.* **36**, 319 (1978).
85. M. J. Fasco, K. P. Vatsis, L. S. Kaminsky, and M. J. Coon, *J. Biol. Chem.*, **253**, 7813 (1978).
86. J. Deutsch, J. C. Leutz, S. K. Yang, H. V. Gelboin, Y. L. Chiang, K. P. Vatsis, and M. J. Coon, *Proc. Natl. Acad. Sci. USA*, **75**, 3123 (1978).
87. R. E. Ebel, D. H. O'Keeffe, and J. A. Peterson, *FEBS Lett.*, **55**, 198 (1975).
88. J. A. Peterson, Y. Ishimura, and B. W. Griffin, *Arch. Biochem. Biophys.*, **149**, 197 (1972).
89. J. O. Stern and J. Peisach, *J. Biol. Chem.*, **249**, 7495 (1974).
90. Y. Miyake, J. L. Gaylor, and H. S. Mason, *J. Biol. Chem.*, **243**, 5788 (1968).
91. J. P. Collman and T. N. Sorrell, *J. Am. Chem. Soc.*, **97**, 4133 (1975).
92. J. O. Stern and J. Peisach, *FEBS Lett.*, **62**, 364 (1976).
93. C. K. Chang and D. Dolphin, *J. Am. Chem. Soc.*, **97**, 5948 (1975).
94. H. H. Ruf and P. Wende, *J. Am. Chem. Soc.*, **99**, 5499 (1977).

95. H. Ogoshi, H. Sugimoto, and Z. Yoshida, *Tetrahedron Lett.*, 2289 (1975).
96. J. P. Collman, T. N. Sorrell, and B. M. Hoffman, *J. Am. Chem. Soc.*, **97**, 913 (1975).
97. S. Koch, S. C. Tang, R. H. Holm, R. B. Frankel, and J. A. Ibers, *J. Am. Chem. Soc.*, **97**, 916 (1975).
98. S. C. Tang, S. Koch, G. C. Papaefthymiou, S. Foner, R. B. Frankel, J. A. Ibers, and R. H. Holm, *J. Am. Chem. Soc.*, **98**, 2414 (1976).
99. V. Ullrich, H. H. Ruf, and P. Wende, *Croat. Chem. Acta*, **49**, 213 (1977).
100. C. K. Chang and D. Dolphin, *Proc. Natl. Acad. Sci USA*, **73**, 3338 (1976).
101. L. K. Hanson, W. A. Eaton, S. G. Sligar, I. C. Gunsalus, M. Gouterman, and C. R. Connel, *J. Am. Chem. Soc.*, **98**, 2672 (1976).
102. C. K. Chang and D. Dolphin, *J. Am. Chem. Soc.*, **98**, 1607 (1976).
103. J. P. Collman, *Acc. Chem. Res.*, **10**, 265 (1977).
104. I. Salmeen and G. Palmer, *J. Chem. Phys.*, **48**, 2049 (1968).
105. M. Sharrock, P. G. Debrunner, C. Schultz, J. D. Lipscomb, V. Marshall, and I. C. Gunsalus, *Biochim. Biophys. Acta*, **420**, 8 (1976).
106. L. Bullard, R. M. Panayappan, A. N. Thorpe, P. Hambright, and G. Ng, *Bioinorg. Chem.*, **3**, 161 (1974).
107. G. Lang, *Q. Rev. Biophys.*, **3**, 1 (1970).
108. M. R. C. Winter, C. E. Johnson, G. Lang, and R. J. P. Williams, *Biochim. Biophys. Acta*, **263**, 515 (1972).
109. U. Gonser and R. W. Grant, *Biophys. J.*, **5**, 823 (1965).
110. K. Ruckpaul, H. Rein, G.-R. Jänig, W. Winkler, and O. Ristau, *Croat. Chem. Acta*, **49**, 339 (1977).
111. J. A. Peterson, *Arch. Biochem. Biophys.*, **144**, 678 (1971).
112. T. Shimizu, T. Nozawa, M. Hatano, Y. Imai, and R. Sato, *Biochemistry*, **14**, 4172 (1975).
113. J. P. Collman, T. N. Sorrell, J. H. Dawson, J. R. Trudell, E. Bunnenberg, and C. Djerassi, *Proc. Natl. Acad. Sci. USA*, **73**, 6 (1976).
114. J. H. Dawson, J. R. Trudell, G. Barth, R. E. Linder, E. Bunnenberg, C. Djerassi, M. Gouterman, C. R. Connell, and P. Sayer, *J. Am. Chem. Soc.*, **99**, 641 (1977).
115. J. H. Dawson, R. H. Holm, J. R. Trudell, G. Barth, R. E. Linder, E. Bunnenberg, C. Djerassi, and S. C. Tang, *J. Am. Chem. Soc.*, **98**, 3707 (1976).
116. B. W. Griffin and J. A. Peterson, *J. Biol. Chem.*, **250**, 6445 (1975).
117. S Maričić, S. Vuk-Pavlović, B. Benko, J. Porok, H. Rein, G.-R. Jänig, and K. Ruckpaul, *Croat. Chem. Acta*, **49**, 323 (1977).
118. R. F. Novak, I. M. Kapetanovic, and J. J. Mieyal, in *Microsomes and Drug Oxidation*, V. Ullrich, I. Roots, A. G. Hildebrandt, R. W. Estabrook, and A. H. Conney, Eds., Pergamon Press, Oxford, 1977, p. 232.
119. H. Rein, S. Böhm, G.-R. Jänig, and K. Ruckpaul, *Croat. Chem. Acta*, **49**, 333 (1977).
120. Y. Ozaki, T. Kitagawa, Y. Kyogoku, H. Shimada, T. Iizuka, and Y. Ishimura, *J. Biochem.*, **80**, 1447 (1976).
121. P. M. Champion and I. C. Gunsalus, *J. Am. Chem. Soc.*, **99**, 2000 (1977).
122. J. A. Peterson, R. E. Ebel, D. H. O'Keeffe, T. Matsubara, and R. W. Estabrook, *J. Biol. Chem.*, **251**, 4010 (1976).

123. J. B. Schenkman, H. Remmer, and R. W. Estabrook, *Mol. Pharmacol.*, **3**, 113 (1967).
124. C. R. Jefcoate, *Methods Enzymol.*, **52**, 258 (1978).
125. Y. Imai, R. Sato, and T. Iyanagi, *J. Biochem.*, **82**, 1237 (1977).
126. H. Schleyer, D. Y. Cooper, O. Rosenthal, and P. Cheung, *Croat. Chem. Acta*, **49**, 179 (1977).
127. F. P. Guengerich, D. P. Ballou, and M. J. Coon, *Biochem. Biophys. Res. Commun.*, **70**, 951 (1976).
128. E. Begard, P. Debey, and P. Douzou, *FEBS Lett.*, **75**, 52 (1977).
129. C. A. Tyson, J. D. Lipscomb, and I. C. Gunsalus, *J. Biol. Chem.*, **247**, 5777 (1972).
130. T. C. Pederson, R. H. Austin, and I. C. Gunsalus, in *Microsomes and Drug Oxidations*, V. Ullrich, I. Roots, A. G. Hildebrandt, R. W. Estabrook, and A. H. Conney, Eds., Pergamon Press, Oxford, 1977, p. 275.
131. B. S. Cohen and R. W. Estabrook, *Arch. Biochem. Biophys.*, **143**, 37 (1971).
132. B. S. Cohen and R. W. Estabrook, *Arch. Biochem. Biophys.*, **143**, 54 (1971).
133. B. S. Cohen and R. W. Estabrook, *Arch. Biochem. Biophys.*, **143**, 46 (1971).
134. Y. Imai and R. Sato, *Biochem. Biophys. Res. Commun.*, **75**, 420 (1977).
135. I. Jansson and J. B. Schenkman, *Mol. Pharmacol.*, **9**, 840 (1973).
136. A. Y. H. Lu, S. B. West, M. Vore, D. Ryan, and W. Levin, *J. Biol. Chem.*, **249**, 6701 (1974).
137. P. Strittmatter, M. J. Rogers, and L. Spatz, *J. Biol. Chem.*, **247**, 7188 (1972).
138. J. Werringloer, in *Microsomes and Drug Oxidations*, V. Ullrich, I. Roots, A. G. Hildebrandt, R. W. Estabrook, and A. H. Conney, Eds., Pergamon Press, Oxford, 1977, p. 261.
139. G. A. Hamilton, in O. Hayaishi, Ed., *Molecular Mechanisms of Oxygen Activation*, Academic Press, New York, 1974, p. 405.
140. F. Lichtenberger, W. Nastainczyk, and V. Ullrich, *Biochem. Biophys. Res. Commun.*, **70**, 939 (1976).
141. D. M. Jerina and J. W. Daly, in T. E. King, H. S. Mason, and M. Morrison, Eds., *Oxidases and Related Redox Systems*, University Park Press, Baltimore, 1973, p. 143.
142. D. M. Jerina, *Chem. Technol.* **4**, 120 (1973).
143. A. D. Rahimtula, P. J. O'Brien, E. G. Hrycay, J. A. Peterson, and R. W. Estabrook, *Biochem. Biophys. Res. Commun.*, **60**, 695 (1974).
144. G. R. Schonbaum and S. Lo, *J. Biol. Chem.*, **247**, 3353 (1972).
145. L. P. Hager, D. L. Doubek, R. M. Silverstein, J. H. Hargis, and J. C. Martin, *J. Am. Chem. Soc.*, **94**, 4364 (1972).
146. H. B. Dunford and J. S. Stillman, *Coord. Chem. Rev.*, **19**, 187 (1976).
147. A. A. Akhrem, D. I. Metelitsa, and M. E. Skurko, *Bioorg. Chem.*, **4**, 307 (1975).
148. H. Mimoun and I. S. deRoch, *Tetrahedron*, **31**, 777 (1975).
149. C. A. Sprecher and A. D. Zuberbuhler, *Angew. Chem. Int. Ed. Eng.*, **16**, 189 (1977).
150. C. Jallabert and H. Riviere, *Tetrahedron Lett.*, 1215 (1977).
151. M. J. Y. Chen and J. K. Kochi, *Chem. Commun.*, 204 (1977).
152. J. T. Groves and G. A. McClusky, *J. Am. Chem. Soc.*, **98**, 859 (1976).
153. J. T. Groves and M. Van Der Puy, *J. Am. Chem. Soc.*, **98**, 5290 (1976).

REFERENCES

154. U. Frommer, V. Ullrich, and H. Staudinger, *Hoppe-Seyler's Z. Physiol. Chem.*, **351**, 903 (1970).
155. V. Ullrich, *Hoppe-Seyler's Z. Physiol. Chem.*, **350**, 357 (1969).
156. J. A. Thompson and J. L. Holtzman, *Drug Metab. Disp.*, **2**, 577 (1974).
157. J. T. Groves, G. A. McClusky, R. E. White, and M. J. Coon, *Biochem. Biophys. Res. Commun.*, **81**, 154 (1978).
158. L. M. Hjelmeland, L. Aronow, and J. R. Trudell, *Biochem. Biophys. Res. Commun.*, **76**, 541 (1977).
159. K. B. Wiberg, *Chem. Rev.*, **55**, 713 (1955).
160. E. M. Kosower, *Physical Organic Chemistry*, Wiley, New York, 1968, p. 130.
161. J. March, *Advanced Organic Chemistry*, McGraw-Hill, New York, 1968, p. 136.
162. A. Hori, S. Takamuku, and H. Sakurai, *J. Org. Chem.*, **42**, 2318 (1977).
163. J. Blanck, G. Smettan, G.-R. Jänig, and K. Ruckpaul, *Acta Biol. Med. Germ.*, **35**, 1455 (1976).
164. B. W. Griffin and J. A. Peterson, *Biochemistry*, **11**, 4740 (1972).
165. E. M. Kosower, *Physical Organic Chemistry*, Wiley, New York, 1968, p. 156.
166. E. G. Hrycay and P. J. O'Brien, *Arch. Biochem. Biophys.*, **153**, 480 (1972).
167. C. Chen and C. C. Lin, *Biochim. Biophys. Acta*, **170**, 366 (1968).
168. A. D. Rahimtula and P. J. O'Brien, *Biochem. Biophys. Res. Commun.*, **60**, 440 (1974).
169. J.-Å. Gustafsson, E. G. Hrycay, and L. Ernster, *Arch. Biochem. Biophys.*, **174**, 440 (1976).
170. H. Danielsson and K. Wikvall, *FEBS Lett.*, **66**, 299 (1976).
171. A. A. Akhrem, S. B. Bokut, and D. I. Metelitsa, *Biochem. Biophys. Res. Commun.*, **77**, 20 (1977).
172. J.-Å. Gustafsson and J. Bergman, *FEBS Lett.*, **70**, 276 (1976).
173. S. G. Sligar, B. S. Shastry, and I. C. Gunsalus, in *Microsomes and Drug Oxidations*, V. Ullrich, I. Roots, A. Hildebrandt, R. W. Estabrook, and A. H. Conney, Eds., Pergamon Press, Oxford, 1977, p. 202.
174. M. J. Coon, R. C. Blake II, D. D. Oprian, and D. P. Ballou, *Acta Biol. Med. Germ.*, **38**, 449 (1979).
175. T. Yonetani, *J. Biol. Chem.*, **240**, 4509 (1965).

CHAPTER 3
Mechanisms of Metal-Catalyzed Oxygen Insertion

JOHN T. GROVES

Department of Chemistry, University of Michigan,
Ann Arbor, Michigan

CONTENTS

1 INTRODUCTION, 127
2 REACTIONS OF METAL–PEROXO COMPOUNDS WITH ALKENES, 128
3 REACTIONS OF METAL–OXO COMPOUNDS WITH ALKANES, 134
4 CHEMICAL MODELS OF BIOLOGICAL OXYGEN TRANSFER, 140
5 HYDROXYLATIONS WITH FERROUS ION–HYDROGEN PEROXIDE, 146
6 HYDROXYLATIONS WITH FERROUS IRON–PEROXY ACIDS, 150
7 OXIDATIONS CATALYZED BY METALLOPORPHYRINS, 155
ACKNOWLEDGMENTS, 159
REFERENCES, 159

1 INTRODUCTION

It is generally understood that while the reaction of organic compounds with oxygen is thermodynamically favored, there are significant kinetic barriers inherent in these processes. The triplet electronic ground state of molecular oxygen requires that any adiabatic process involving electron transfer from a stable, closed shell (singlet) organic species to dioxygen result in the initial formation of a triplet diradical product. The important role of metal ions in oxidation processes of biological importance has been recognized for a long time (1), and correspondingly there has been sustained interest in elucidating the molecular details of catalytic and stoichiometric metal-mediated oxidations (2, 3).

Metals which serve as catalysts for the oxidation of organic compounds must, either directly or indirectly, also be catalysts for the reduction of oxygen. Since dioxygen is a potential four-electron acceptor, its interactions with metals of variable valence is expected to be complex. Nonetheless it is instructive to consider five formal stages of metal-to-oxygen electron transfer as outlined in Scheme 1.

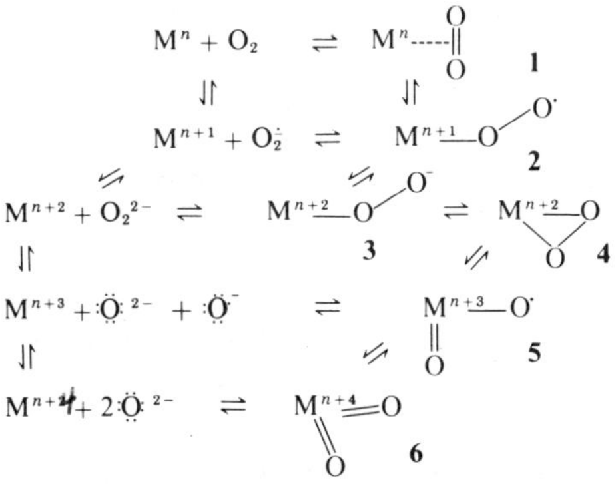

$$M^n + O_2 \rightleftharpoons M^n\text{-----}\overset{O}{\underset{O}{\|}} \quad 1$$

$$\updownarrow \qquad \updownarrow$$

$$M^{n+1} + O_2^{\cdot -} \rightleftharpoons M^{n+1}\text{--}O{\diagdown}_{O^{\cdot}} \quad 2$$

$$M^{n+2} + O_2^{2-} \rightleftharpoons M^{n+2}\text{--}O{\diagdown}_{O^{-}} \rightleftharpoons M^{n+2}{\diagdown \atop O}{\diagup \atop O} \quad 3,\ 4$$

$$\updownarrow$$

$$M^{n+3} + \ddot{O}^{2-} + \cdot\ddot{O}^{-} \rightleftharpoons M^{n+3}\text{--}O^{\cdot} \quad 5$$
$$\qquad\qquad\qquad\qquad \underset{O}{\|}$$

$$\updownarrow$$

$$M^{n+4} + 2\ddot{O}^{2-} \rightleftharpoons M^{n+4}\text{=}O \quad 6$$
$$\qquad\qquad\qquad \overset{\|}{O}$$

Scheme 1

The vertical column of reactions on the left side of Scheme 1 represents the familiar serial reduction of dioxygen by one-electron steps to superoxide ion, peroxide ion, hydroxyl radical, and finally to the oxidation state of water. The column of reactions on the right side depicts the corresponding series of metal–oxygen complexes at the same stages of redox interaction.

Complex **1** is a metal–dioxygen complex in which there has been no net electron transfer from the metal to oxygen. One-electron and two-electron transfers from the metal would lead to a metal–superoxide complex **2** and two possible metal–peroxo complexes **3** and **4**, respectively.

Transfer of a third electron from the metal to the peroxy ligand in **4** requires the rupture of the O—O bond to give a metal–oxy radical species **5**. The metal oxo species **6** is the formal result of the transfer of a fourth electron from the metal to the oxygen ligands. It is possible to consider **5** and **6** as resonance forms of the same structure if they are of the same electronic spin state.

In principal any of the metal species, any of the oxygen species, or any of the metal–oxygen complexes could serve as the primary oxidant in reactions with organic compounds. Indeed there are bona fide examples of each type of reaction. The purpose of this chapter is to survey the range of reaction mechanisms observed for oxygen insertion into organic compounds and to attempt to outline a chemical basis for the notion of oxygen activation.

2 REACTIONS OF METAL–PEROXO COMPOUNDS WITH ALKENES

A large number of stable metal–dioxygen complexes have been synthesized (4, 5). The reactions of these complexes, containing mostly low-valent Group VIII transition metals, have been shown to proceed via free-radical chain reactions involving organic peroxide intermediates (6).

The reactions of molybdenum-, chromium-, platinum-, palladium-, iridium-, and rhodium–peroxo complexes appear to be exceptional in this regard. It has been shown by Whitesides that nucleophilic attack of n-butyllithium will occur with peroxymolybdenum and peroxochromium species but that the peroxo complexes of platinum and iridium are essentially nucleophilic (7).

Sheldon has reported that reaction of palladium(II)- and platinum(II)–peroxo complexes with electron-deficient olefins gives a family of 1:1 adducts for which a peroxo metallacycle structure **7** was suggested on the basis of NMR and IR data (8). A similar platinum–dioxygen complex with tetracyanoethylene has been reported by Ugo (9).

Thermal decomposition of the peroxo metallacycle **7** was reported to give ketones *via* carbon-carbon bond cleavage and a new metal complex tentatively suggested to be a metallaoxirane (**7a**). By contrast, acidic decomposition of **7** caused partitioning to recovered starting material and epoxide (**8**).

It was concluded that these platinum and palladium peroxo species were behaving as nucleophilic reagents that added to these electron-poor olefins in a manner analogous to the well-known addition of basic hydrogen peroxide. A stepwise process involving the zwitterionic intermediate **9** was proposed:

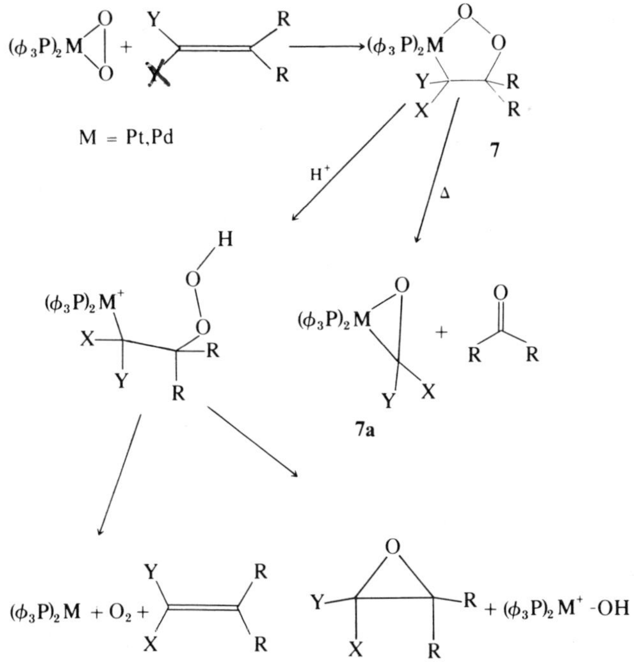

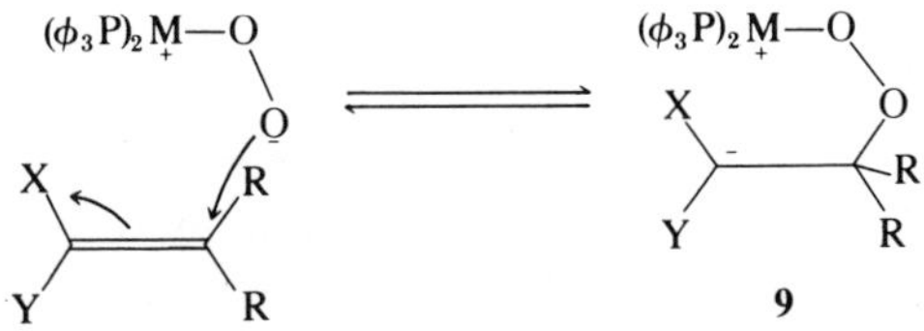

Peroxo species of tungsten, vanadium, and molybdenum are known to be effective epoxidation reagents. Thus treatment of maleic acid with hydrogen peroxide in the presence of sodium tungstate or vanadium pentoxide gives a good yield of the corresponding epoxide (10). The mechanism is thought to involve the corresponding peroxytungstic and peroxyvanadic acids (11):

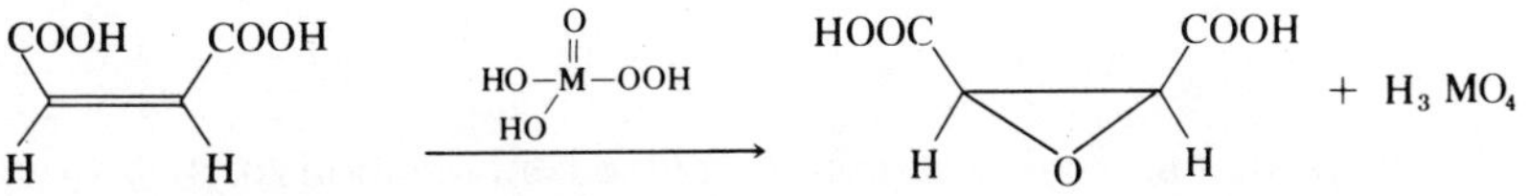

$H_2O_2 + H_3MO_4 \rightleftharpoons (HO)_2M\text{--}OOH + H_2O$

$M = W, V$

Mimoun has shown that molybdenum(V) peroxo compounds are effective epoxidizing reagents (12). The mechanism originally proposed for this reaction involved initial π complexation of the olefin to molybdenum and insertion of the π bond into a metal–oxygen bond to give a peroxometallacycle **10**:

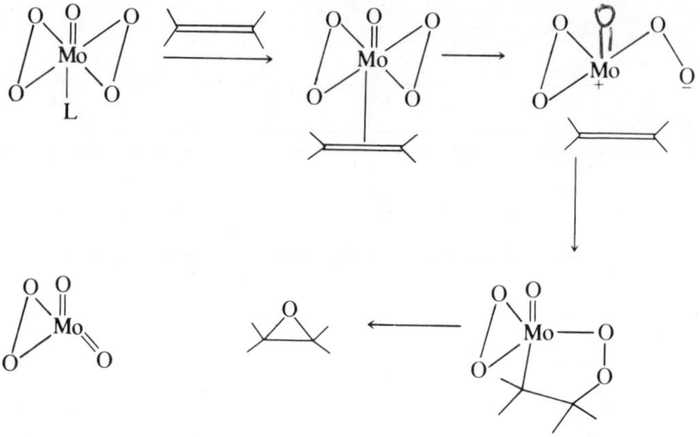

Subsequently it was demonstrated by Sharpless that an ^{18}O label in the metal–oxo ligand was not significantly incorporated into the product epoxide and that the peroxo oxygen was involved in the oxygen transfer step.[13] It was noted however that relative olefin reactivities were very much unlike those expected for a 1,3-dipolar cycloaddition reaction and resembled rather reactivities observed for peroxy acids. On this basis transition states **11** and **12** were suggested as more likely alternatives. Similar arguments have been advanced by Roček in support of a symmetrical, three-membered transition state for the olefin epoxidation reactions of chromic acid and related reagents (14).

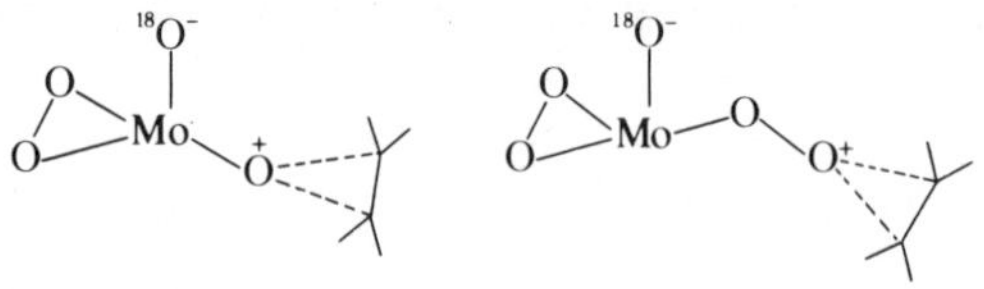

The molybdenum- and vanadium-catalyzed epoxidations of olefins with alkyl hydroperoxides apparently do not involve regeneration of MoO_5 as the oxidizing agent. Sharpless has shown that ^{18}O in the peroxo ligands of the metal is not incorporated into the product epoxide (15). It was proposed that the reactive species was a metal–peroxo complex. The exceptional reactivity of allylic alcohols in this reaction can then be adequately explained as being due to ternary complexation.

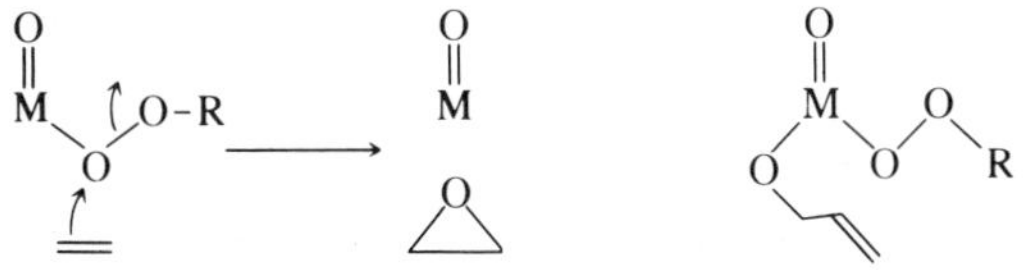

It has been reported by Reed (16) and Milner (17) that rhodium(I) complexes can effect the conversion of terminal olefins to methyl ketones in the presence of oxygen. Two types of mechanisms have been proposed for this process: (a) a Wacker-type electrophilic addition of Rh^{III}—OH to the olefin and (b) reaction of a rhodium–oxygen complex with the olefin to give a peroxo metallacycle (13) analogous to 10 above:

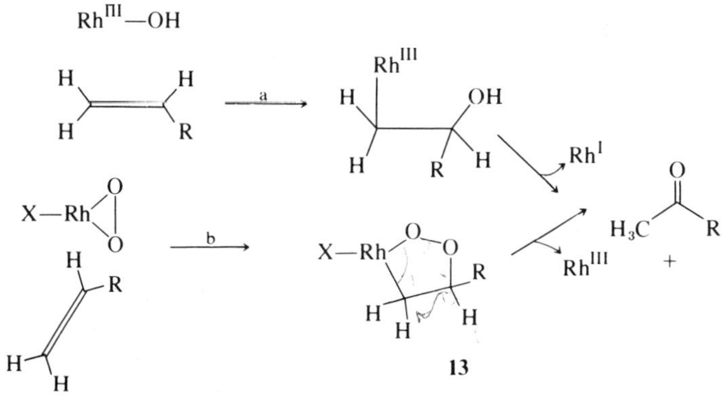

In the former case the oxygen should derive exclusively from water in the medium, whereas in the latter case only oxygen from O_2 will be incorporated into the product ketone.

Mimoun has proposed a coupling of these two processes in which the Rh(I) produced as a result of the Wacker cycle binds and activates molecular oxygen via the peroxo metallacycle route (Scheme 2) (18). It was suggested that the oxidative cleavage of the double bond that accompanies ketone formation was due to competing carbon–carbon bond cleavage in this peroxo metallacycle 14. Alternatively it was suggested that the Rh(III) oxo complex 15 effected this oxidative cleavage in a manner reminiscent of the olefin metathesis reaction.

In further support of this scheme, a Rh(III) peroxo metallacycle (16) from the reaction of a rhodium–oxygen complex with tetracyanoethylene

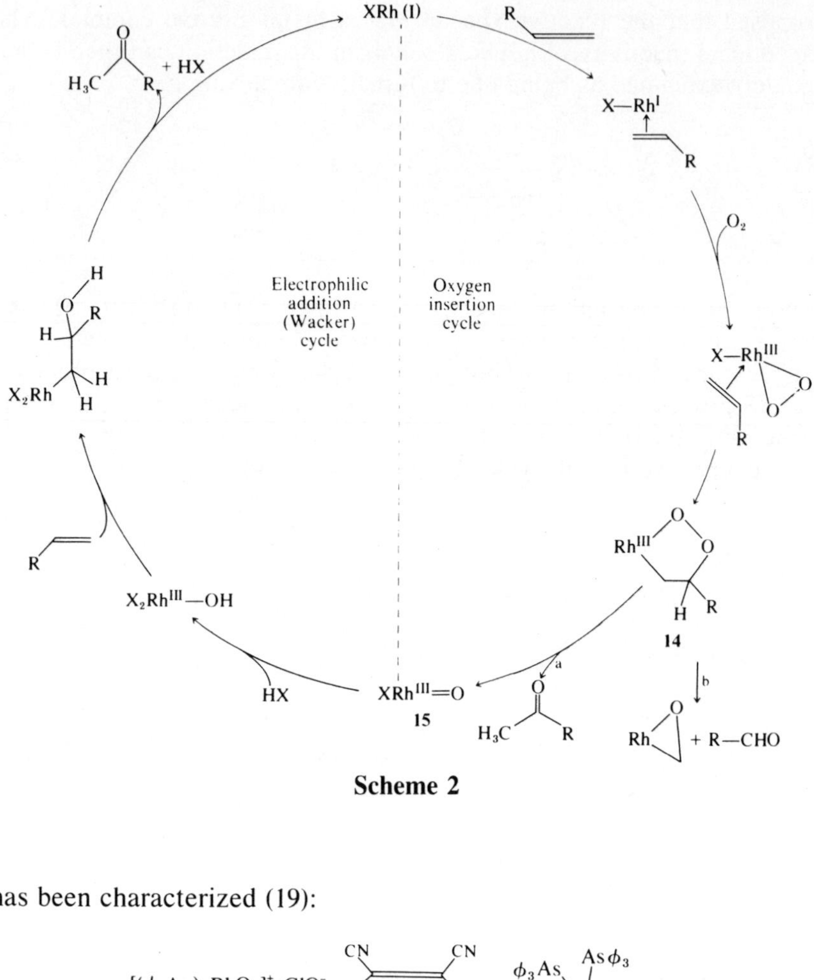

Scheme 2

has been characterized (19):

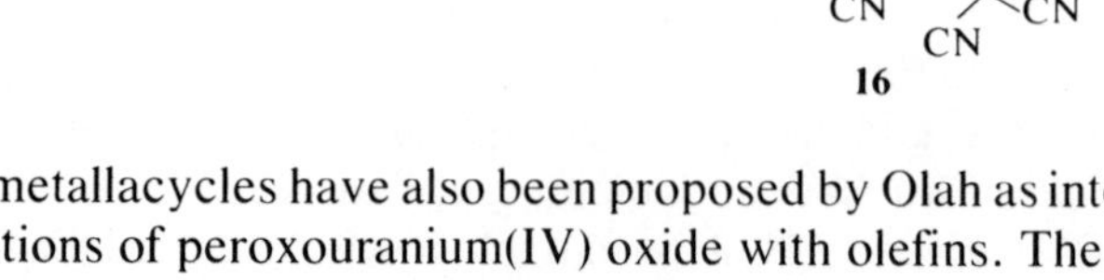

$$[(\phi_3As)_4RhO_2]^+ \ ClO_4^-$$

followed by reaction with tetracyanoethylene (NC-C(CN)=C(CN)-CN) to give **16**.

Peroxo metallacycles have also been proposed by Olah as intermediates in the reactions of peroxouranium(IV) oxide with olefins. The mixture of epoxide **17** and rearrangement products **18** was rationalized by initial addition to the olefin to the coordinately unsaturated UO_4 to give **19** and subsequent ring opening to give a zwitterionic intermediate **20**:

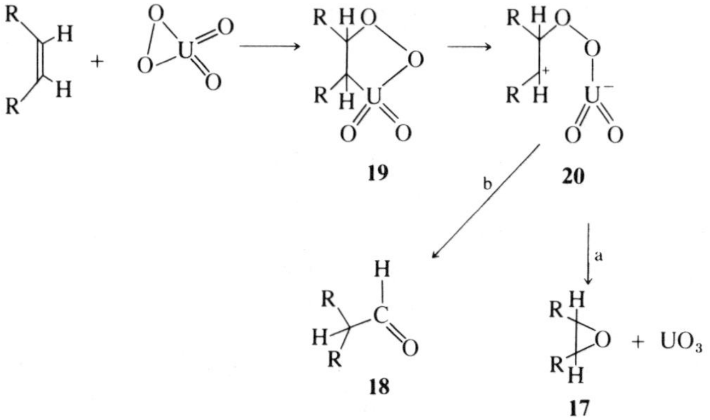

A novel epoxidation reaction which involves the reaction of platinum–dioxygen complexes with benzoyl chloride has been reported by Kochi (21). Thus treatment of $(\phi_3P)_2PtO_2$ with benzoyl chloride in the presence of an olefin led to the formation of the corresponding epoxides. The reaction of **21** with benzoyl chloride in the absence of olefin gave a Pt(II)–peroxy acid complex (**22**). This complex was shown to be a reactive oxygen species even at $-78°C$:

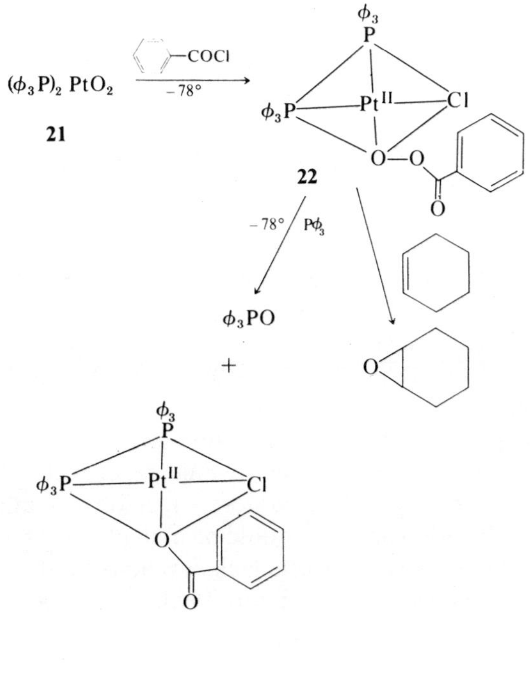

3 REACTIONS OF METAL–OXO COMPOUNDS WITH ALKANES

The oxidations of organic compounds by various transition metal–oxo compounds constitute an important class of oxygen insertion reactions. Alkanes are hydroxylated by such reagents to give alcohols, and alcohols lead to ketones, aldehydes afford carboxylic acids, and alkenes undergo allylic oxidation, epoxidation, vicinal oxidation, and cleavage. Extensive mechanistic investigations of these reactions indicate that the oxo ligands and perhaps also the metal center are intimately involved in these processes (22, 23).

For the oxidation of alkanes by chromic acid, a detailed mechanistic picture has emerged. The reaction is first order in alkane, CrO_3, and hydrogen ion, and the relative reactivities of primary, secondary, and tertiary hydrogens are 1:110:7000 (24). A hydrogen isotope effect of circa 2.5 has been observed for tertiary hydrogens (25). The relative rates of cycloalkane oxidation and the lack of reactivity of bridgehead hydrogens indicate a changing geometry, tetrahedral to trigonal, in the transition state of these hydroxylations (26). This fact is also reflected by the observation that sterically hindered tertiary centers are oxidized more rapidly than less hindered sites.

Wiberg and Foster have observed that the oxidation of a chiral tertiary alkane, (+)-3-methyl heptane (**23**), gave alcohol **24** with 72 to 85% retention of configuration at the oxidized carbon. Further, alcohols are the primary products even in acetic acid solvent (25). Thus the new hydroxyl group derived from the oxo ligand of chromium.

Wiberg has considered three processes by which the C–H bond cleavage can take place: (a) hydride abstraction by Cr(VI) to give a carbocation (**25**) and Cr(IV), (b) hydrogen atom abstraction to give a carbon radical (**26**) and Cr(V), and (c) direct C–H insertion by Cr(VI) to give a chromate ester (**27**) without intermediates (22).

The hydride abstraction path appears unlikely on the grounds that substituent groups such as phenyl which would be expected to greatly stabilize a carbocation had only a modest effect on the rate of reaction. Likewise the C–H insertion route would be unlikely to exhibit accelerated rates at sterically hindered sites or a lack of reactivity at bridgehead centers. For these reasons the best picture of the C–H bond cleavage step

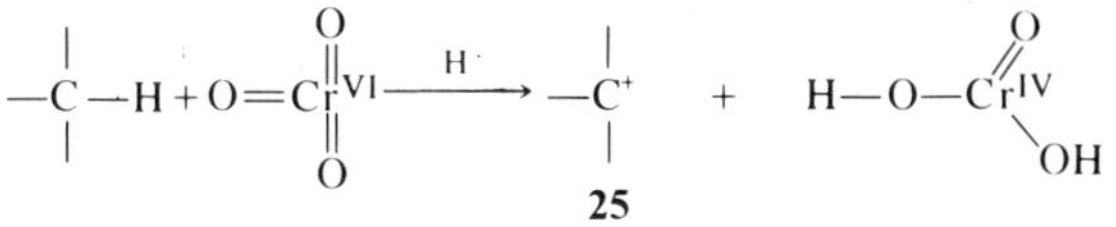

$$-\underset{|}{\overset{|}{C}}-H + O=\underset{\underset{O}{\|}}{\overset{O}{\|}}Cr^{VI} \xrightarrow{H\cdot} -\underset{|}{\overset{|}{C}}^{+} \ + \ H-O-\underset{OH}{\overset{\overset{O}{\|}}{Cr^{IV}}} \quad (a)$$

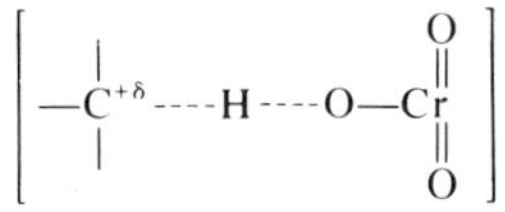

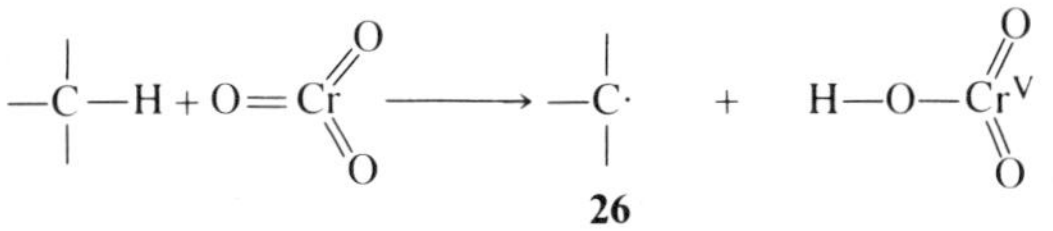

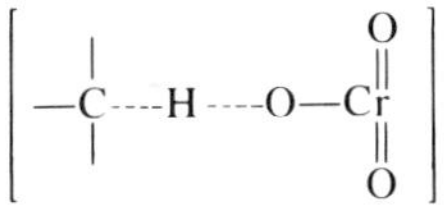

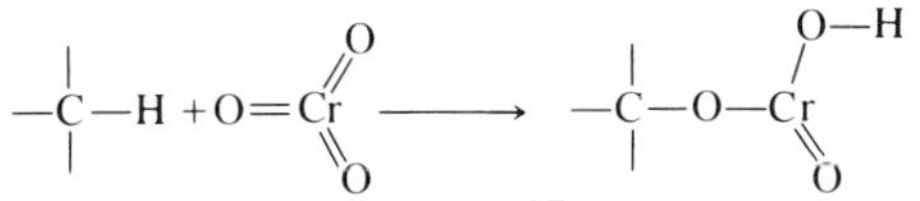

$$\left[-\underset{|}{\overset{|}{C}}^{+\delta}----H----O-\underset{\underset{O}{\|}}{\overset{O}{\|}}Cr \right]$$

$$-\underset{|}{\overset{|}{C}}-H + O=\underset{\underset{O}{\diagdown}}{\overset{O}{\diagup}}Cr \longrightarrow -\underset{|}{\overset{|}{C}}\cdot \ + \ H-O-\underset{\underset{O}{\diagdown}}{\overset{O}{\diagup}}Cr^{V} \quad (b)$$

26

$$\left[-\underset{|}{\overset{|}{C}}----H----O-\underset{\underset{O}{\|}}{\overset{O}{\|}}Cr \right]$$

$$-\underset{|}{\overset{|}{C}}-H + O=\underset{\underset{O}{\diagdown}}{\overset{O}{\diagup}}Cr \longrightarrow -\underset{|}{\overset{|}{C}}-O-\underset{\underset{O}{\diagdown}}{\overset{O-H}{\diagup}}Cr \quad (c)$$

27

$$\underset{\diagup}{\overset{\diagdown}{C}}-----H$$
$$O \diagdown \diagup O$$
$$\underset{\overset{\|}{O}}{Cr}$$

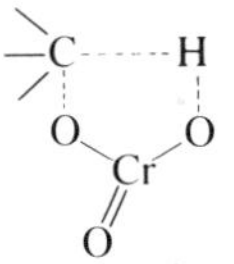

in chromate oxidation appears to be a hydrogen atom abstraction reaction rather similar to a free-radical bromination.

More recently evidence has emerged in favor of a similar hydrogen atom abstraction in the oxidation of aldehydes by chromyl acetate (**27**). In this case decarbonylation of the intermediate acyl radical (**28**) was observed when R was triphenylmethyl and was able to stabilize a carbon radical at the benzylic carbon. It is unlikely that such a decarbonylation reaction would occur by the alternative two-electron process that has been suggested for this reaction.

Retention of configuration at carbon in a free-radical hydrogen abstraction requires that the incipient radical not be freely rotating. This can be accommodated by rapid collapse of a free-radical cage structure to give an alkyl chromate ester. Such a process is consistent with the observation that nucleophilic solvents do not generally compete with the oxo ligand of the metal.

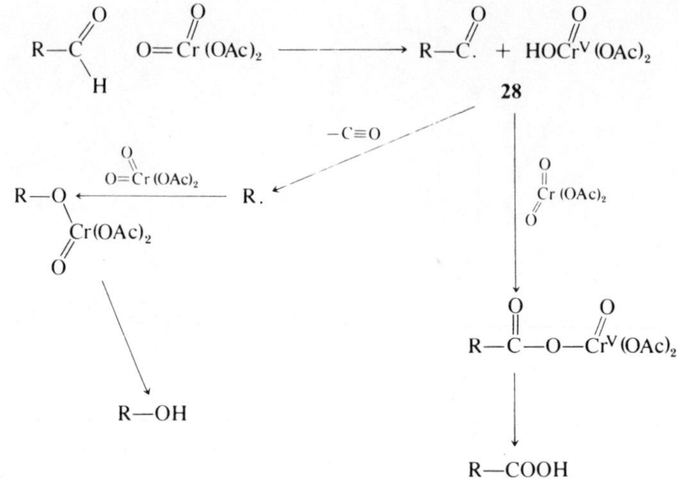

$$\underset{H}{\overset{O}{R-C}} \quad O=\overset{O}{Cr}(OAc)_2 \longrightarrow R-\overset{O}{C}\cdot + HO\overset{V}{Cr}(OAc)_2$$

$$\mathbf{28}$$

$$R-O \xleftarrow{\overset{O=Cr(OAc)_2}{}} R\cdot \xleftarrow{-C\equiv O}$$

$$\underset{O}{\overset{Cr(OAc)_2}{}} \quad \overset{O}{\underset{O}{Cr(OAc)_2}}$$

$$R-\overset{O}{C}-O-\overset{V}{Cr}(OAc)_2$$

$$R-OH \qquad R-COOH$$

The oxidation of radical centers by metal ions has been examined in considerable detail by Kochi (28). There is little reason to doubt that carbon radicals generated by hydrogen abstraction will not behave similarly to radicals generated by other processes. In general one can consider (a) an outer-sphere electron transfer process to give a carbocation and a reduced metal ion without perturbing the solvation sphere of the metal, (b) addition of the radical to the metal to give an organometallic intermediate, and (c) homolytic displacement of a ligand to effect ligand transfer from the metal to carbon. The metal–alkyl intermediate can subsequently react to give products of either ligand transfer, electron transfer, or other products:

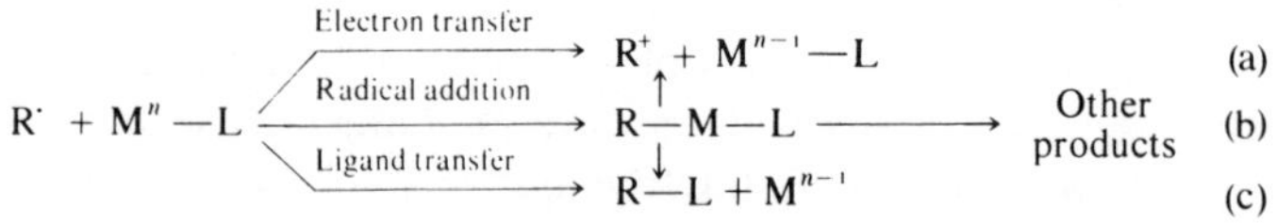

$$R\cdot + M^n-L \begin{array}{c} \xrightarrow{\text{Electron transfer}} R^+ + M^{n-1}-L \\ \xrightarrow{\text{Radical addition}} R-M-L \\ \xrightarrow{\text{Ligand transfer}} R-L + M^{n-1} \end{array} \xrightarrow{} \text{Other products} \begin{array}{c} (a) \\ (b) \\ (c) \end{array}$$

The products of electron transfer oxidation of radicals are similar to those of other processes such as solvolysis reactions which generate cationic intermediates. Thus for example, the electron transfer oxidation of carbon radicals generated in water would lead to alkenes by loss of a proton and alcohols by nucleophilic attack of solvent. Typical cationic rearrangements of the carbon skeleton would also be expected.

The oxidation of cyclobutyl radicals (**29**) by lead tetraacetate in acetic acid gives a distribution of acetate products very similar to that observed for the solvolysis of cyclobutyl tosylate **30** (29):

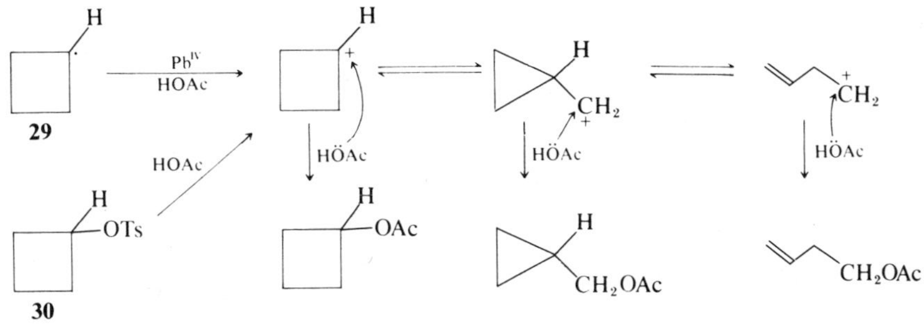

REACTIONS OF METAL–OXO COMPOUNDS WITH ALKANES

By contrast the ligand transfer process of radical oxidation does not involve discreet carbocation intermediates and rearrangements are not normally observed in these processes. The reaction can be conceived to pass through a bridged inner-sphere transition state typical of any free-radical atom transfer reaction. The nature of the metal, the carbon radical, and the ligand all influence the facility of ligand transfer.

Cupric bromide has been shown to be particularly facile at capturing carbon free radicals by ligand transfer. Near diffusion-controlled rates of bromide transfer are inferred from the observation that even the cyclopropyl carbinyl radical **31** which is known to ring-open with a rate of 10^8/sec is captured without rearrangement (30):

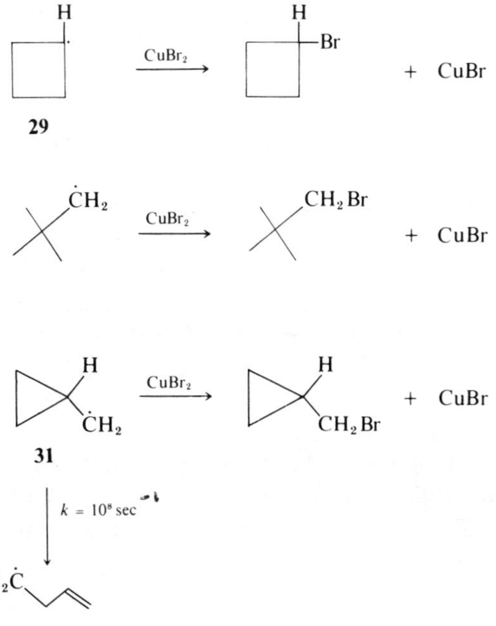

Kochi has suggested that hard nucleophiles such as oxy anions are relatively inefficient in ligand transfer processes, in contrast to the softer halides and pseudohalides. With hard ligands addition of the radical to the metal center to give a metal–alkyl intermediate is preferred. Substitution at carbon can then proceed by what amounts to a reductive elimination of the carbon ligand (28).

Applied to the hydroxylation of alkanes by chromium–oxo reagents, the initially formed Cr(V) carbon-free radical cage structure 32 could collapse by radical attack on oxygen to give a Cr(IV) ester (33) directly. Alternatively radical addition to the metal could give rise to an alkyl chromium intermediate (34) which decomposes by carbon migration from chromium to oxygen. Such a migration would be expected to occur with retention of configuration at carbon:

$$\begin{array}{c}\text{C-H} \quad \text{O=Cr}^{VI}\text{-X}_2 \longrightarrow \left[\text{C·} \quad \text{HO-Cr}^{V}\text{X}_2 \right] \\ \swarrow \qquad\qquad 32 \qquad\qquad \searrow \\ \text{C-Cr}^{VI}\text{X}_2\text{(OH)} \longrightarrow \text{C-O-Cr}^{IV}\text{X}_2\text{(OH)} \\ 34 \qquad\qquad\qquad 33 \end{array}$$

There appears to be insufficient data at present to differentiate these two paths; but since Cr(V) is a d_1 metal ion, rapid collapse of such a radical *pair* is quite reasonable.

Sharpless has proposed that the intermediacy of organometallic intermediates may be a general phenomenon in the oxidation of hydrocarbons by metal–oxo compounds (32). The reaction of chromyl chloride with *cis*-alkenes at low temperature was found to give largely *cis*-epoxide, as well as chlorohydrin and dichloride resulting from cis addition to the double bond. This unusual cis addition was used to support a mechanism in which an initial olefin–chromium π complex (35) rearranged to form a four-membered metallacycle (36) with retention of relative configuration of the carbon substituents. Subsequent chromium to ligand rearrangement, analogous to those described above, could lead to the observed *cis*-epoxide 37 or *cis*-chlorohydrin 38. The *cis*-dichloride was considered to arise from insertion of the olefin into a chromium–chlorine bond (Scheme 3). It was noted that *cis*-addition could also arise from attack of the two olefinic carbons on the metal ligands in a 5-membered transition state 39.

The hydroxylation of aliphatic centers by manganese–oxo (23) and vanadium–oxo (33) compounds has been found to proceed in a manner very

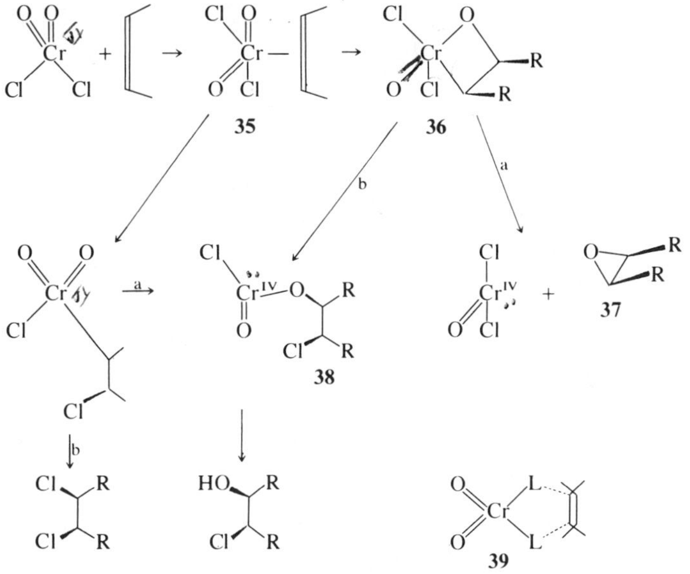

Scheme 3

similar to that of chromic acid. Thus Brauman has found that tertiary benzylic sites are hydroxylated by basic permanganate with a very large isotope effect (k_H/k_D = 11.5) and a significant degree of retention of configuration at the oxidized carbon (34). Collapse of a radical cage structure (**40**) with ultimate transfer of the metal–oxo ligand to carbon was proposed here as well:

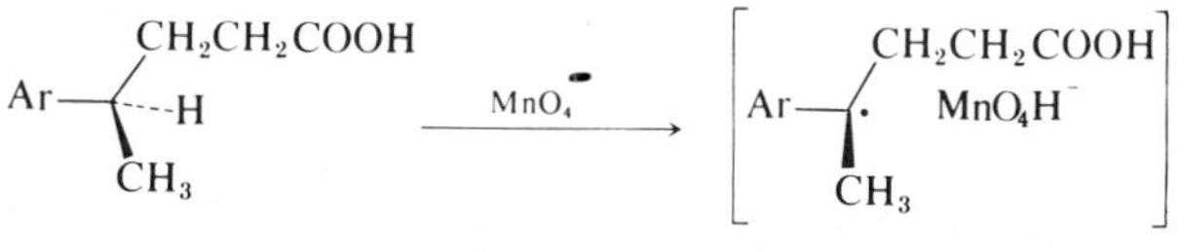

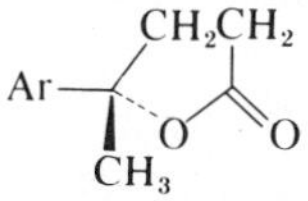

Rather little chemistry of higher oxidation states of iron has been reported. Alcohols are known to be oxidized to aldehydes or ketones by potassium ferrate (35–37) and squalene has been reported to be oxidized in low yield by barium ferrate (38). Because such oxygen transfer reactions can be considered as prototypes of the reactions of biological oxygenases, we have examined the oxidation of alkanes by ferrate(VI) salts.

The decomposition of K_2FeO_4 in water at pH $\approx$ 9 has been reported to be second order in FeO_4^{2-} and first order in H^+ and apparently produces FeO_4^{4-} in concentrated NaOH. The data support a decomposition scheme which produces a transient Fe(IV) species. Although K_2FeO_4 is unreactive toward alkanes by itself, the decomposition of K_2FeO_4 by oxalic acid in the presence of alkanes leads to hydroxylation (39). Adamantane was oxidized under these conditions to give 1-adamantanol and 2-adamantanone in a ratio of 12:1. Only alcohols were produced when the reaction was run in glacial acetic acid, indicating that this reaction involves transfer of the oxo ligand to carbon (40). Acetamides were produced in the same isomeric ratio when the reaction was run in acetonitrile. The formation of acetamides is best rationalized as the result of the alkylation of nitrogen by intermediate carbocations.

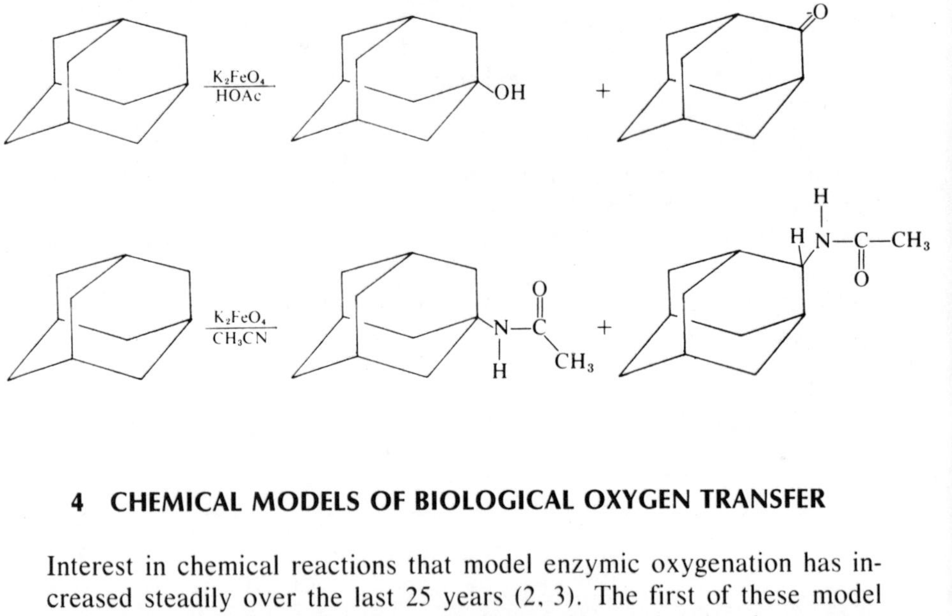

4 CHEMICAL MODELS OF BIOLOGICAL OXYGEN TRANSFER

Interest in chemical reactions that model enzymic oxygenation has increased steadily over the last 25 years (2, 3). The first of these model reactions, a mixture of ascorbic acid, ferrous ion, and molecular oxygen, has come to be known as the Udenfriend system (41). This reagent has been shown to oxidize aromatic hydrocarbons to give phenols, to hydroxylate aliphatic hydrocarbons, and to epoxidize olefins.

A related family of similar oxidizing systems has emerged with diaminopurine (42), catechol (43, 44), mercaptobenzoic acid (45), tetrahydrobiopterine (46), N-benzyl-1,4-dihydronicotinamide (47), and hydrazobenzene (48) as the reducing agent. The various mechanisms proposed for these systems have been recently reviewed (49).

Perhaps surprisingly, and despite the many thorough studies of the nature of these reactions, no simple picture of their mechanisms has emerged. Many of these studies have relied heavily on comparisons of product distributions, particularly in aromatic hydroxylation. A serious problem in interpreting results of this kind is that there is no guarantee that the product distribution accurately reflects the initial point of attack or the nature of the oxidant. Walling has clearly demonstrated that wide variations in isomer ratios can be obtained under conditions in which hydroxyl radical is expected to be the primary oxidant by changing the nature and concentration of the metal ion (50).

Some of the mechanistic complexities of the Udenfriend system (41) and the ferrous ion–mercaptobenzoic acid oxygen system of Ullrich (45) have been examined recently by Lindsay Smith et al. (51). The oxidation of naphthalene in acetone with oxygen, ferrous ion, and mercaptobenzoic acid gave α- and β-naphthol (**41** and **42**), cis- and trans-1,2-dihydroxy-dihydronaphthalene (**43** and **44**), and several other more highly oxidized products:

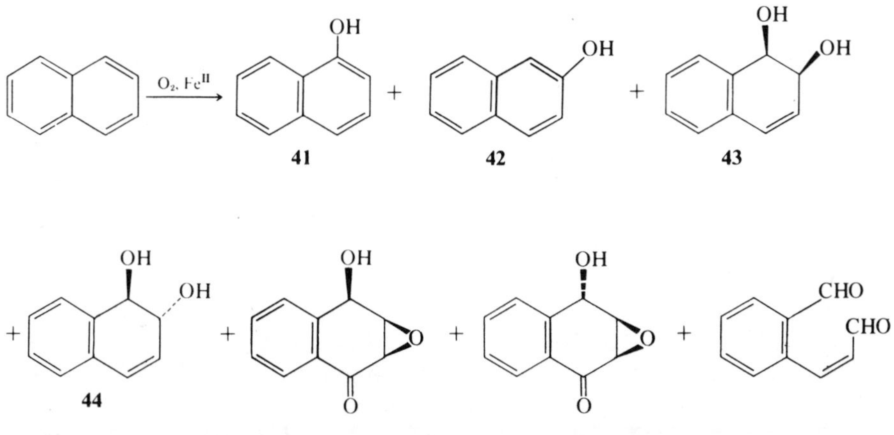

No NIH shift of hydrogen was associated with the production of **42**, while a maximum of 11% of the label had migrated to the adjacent carbon during the formation of **41**.

When the reaction was run in the presence of $^{18}O_2$ and $^{16}O_2$, the isotopic content of the product diols indicated that *both* oxygen atoms had come from molecular oxygen but from different oxygen molecules. This result and the mixture of stereoisomers appear to rule out epoxides or oxetanes as intermediates in this reaction.

The oxidation of naphthalene by Fe(II), EDTA, ascorbic acid, and oxygen (the Udenfriend system) gave a similar mixture of products. Interestingly the ratio of naphthols to dihydrodiols was found not to depend on the presence or absence of ascorbic acid. This observation would tend to limit the importance of ascorbic acid as an integral part of the primary oxidant as has been suggested by Hamilton (44). More likely the ascorbic acid is serving to maintain the iron in the ferrous state. The relative yields of 1- and 2-naphthol varied with Fe(II) concentration and with deuteration of the ring.

The authors conclude, reasonably, that the Udenfriend system and the related Ullrich system are metal-catalyzed free-radical processes that do not pass through an arene oxide and thus are mechanistically distinct from the biological oxygenases. The reactive oxygen species was proposed to be a metal ion–oxygen complex which acts as a radical oxidant. The initial phases of the mechanism are proposed to be as shown in Scheme 4:

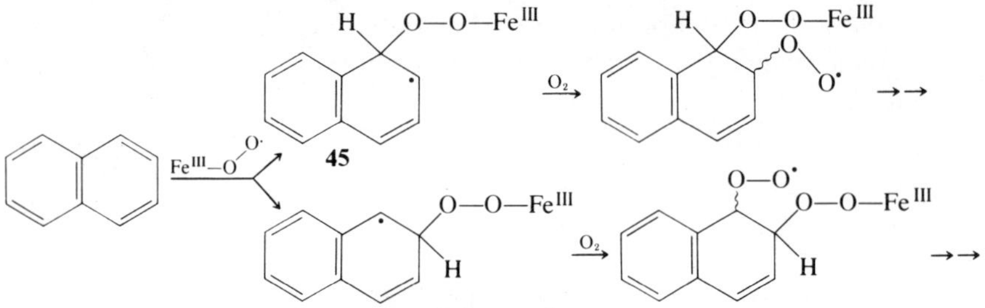

Scheme 4

Such a complex has also been proposed by Hamilton (3) to be involved in similar oxygenations but there is at present no firm evidence that an iron–superoxide complex will add to double bonds in this manner.

Support for intermediate iron–peroxy complexes such as **45** can be derived from the related aromatic oxidation by oxygen in the presence of Co(II) complexes. Nishinaga has shown that para-substituted 2,6-di-*t*-butylphenols (**46**) react with Co(II)salpr [salpr = bis(3-salicylideneaminopropyl-amine] (**47**) and oxygen to give an isolable Co(III)–peroxo complex **48** (52):

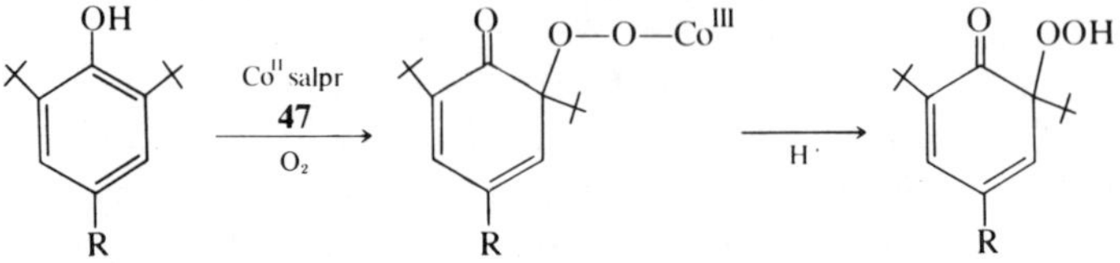

Filtration of the peroxy–Co(III) complex through silica gel allowed the isolation of the stable ketohydroperoxide **49**. The mechanism of this transformation has been examined by ESR. Treatment of Co(II) salpr with oxygen led to the appearance of an eight-line pattern ($A_{Co} = 136$) typical of a 1:1 Co–O$_2$ complex. Treatment of this complex with trisubstituted phenols caused a disappearance of these signals and the appearance of a phenoxy radical pattern. The authors proposed that the intermediate phenoxy radical was reduced by Co(II) to give a Co(III) phenoxide (**50**) which spontaneously added oxygen (Scheme 5). Molecular oxygen reacted slowly with the phenoxy radical and the reaction of the phenoxy radical directly with Co(III)O$_2^{2-}$ salpr gave para-substituted peroxides, apparently ruling out these otherwise more reasonable alternatives.

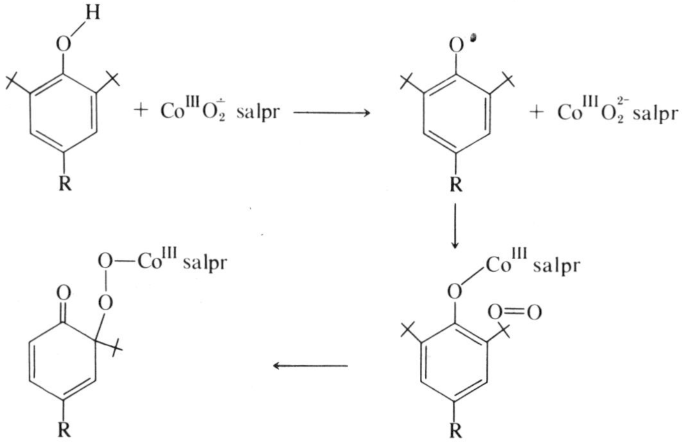

Scheme 5

Cobalt(II) complexes of bis(salicylidene)ethylenediamine (salen) have been shown to cause the oxidative cleavage of indole derivatives (**51**) in the presence of oxygen (63):

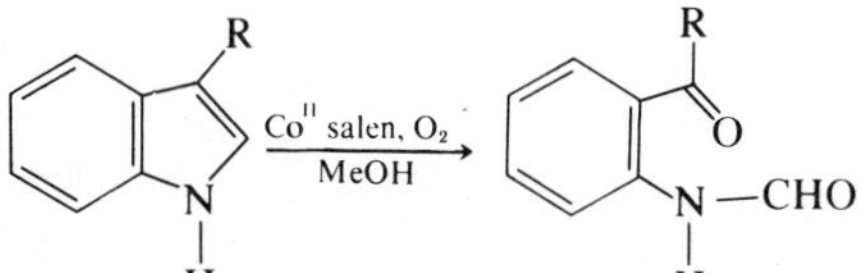

This reaction is of considerable interest since it models the reactions of indoleamine-2,3-dioxygenase (54–56). Interestingly, the relative reactivity of various indoles correlated with the electron donor properties of the indole. Indole-3-acetic esters were found to give side-chain methoxylation in methanol. Taken together these results suggest that the mechanism of this reaction may be initial electron transfer to give an indole radical cation (**52**). Reaction of this radical cation with Co(III)–O_2^- or oxygen followed by Co(II) reduction would lead to a peroxy indole Co(III) complex (**53**) analogous to that observed in the related phenol oxidations.

The side-chain methoxylation could be readily explained by oxidation of the initially formed radical cation **52** to give **54**, a process well precedented for the Co(III) oxidation of toluene (57) and subsequent addition of methanol (Scheme 6):

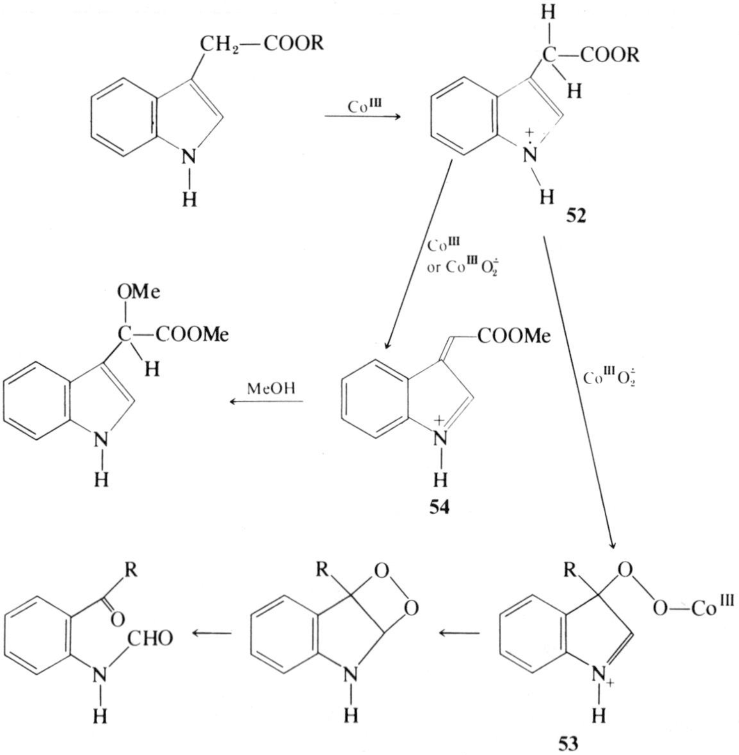

Scheme 6

The direct addition of oxygen to π-radical cations has recently been shown to lead to endoperoxides (**55** and **56**). The reaction of dienes with oxygen in the presence of catalytic amounts of triarylamine radical cations was first observed by Barton (58, 59). It was initially suggested that the role of the catalyst was to affect the spin-forbiddenness of a direct reaction between ground state $^3O_2$ and a singlet diene by complex formation. It was subsequently shown that $VOCl_3$, $MoCl_5$, $FeCl_3$, and aromatic hydrocarbon radical cations were also effective (60).

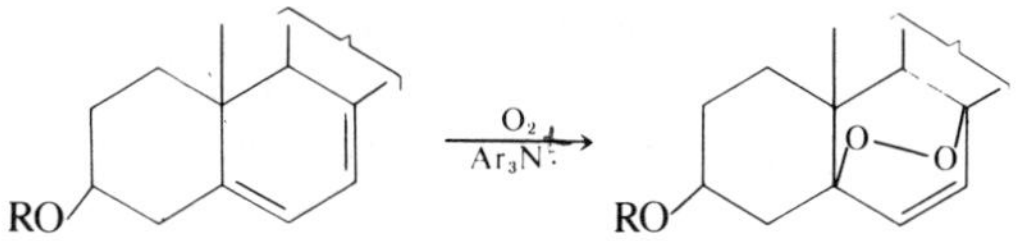

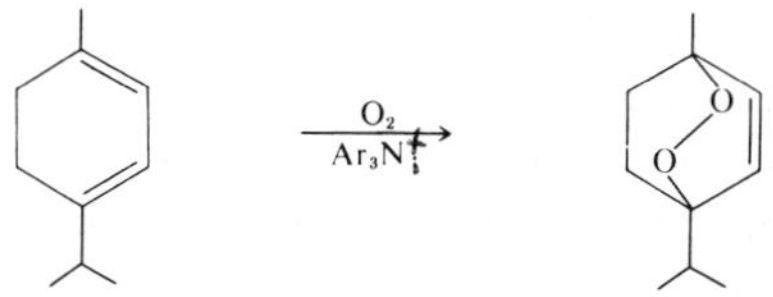

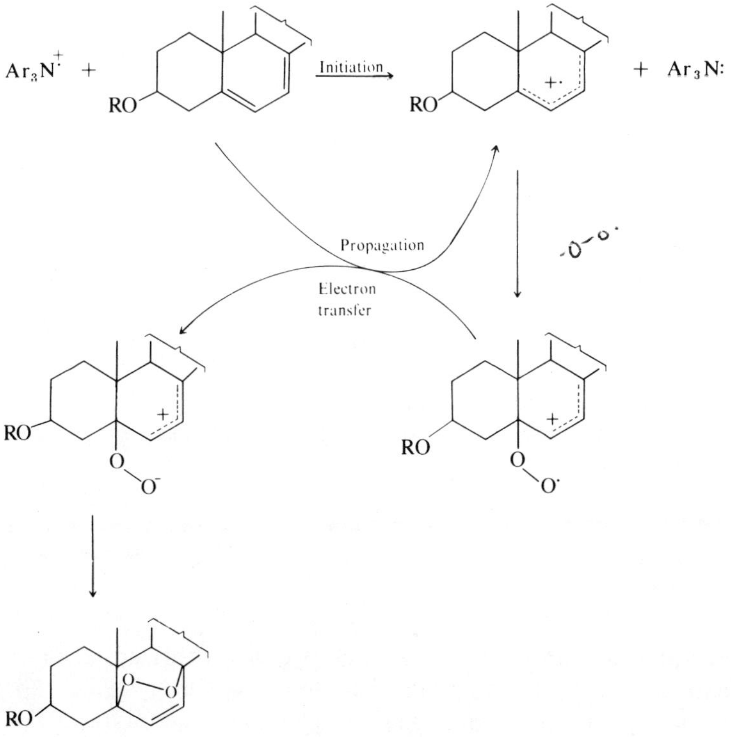

Scheme 7

It has been convincingly demonstrated that the mechanism of this unusual process is a radical-cation chain reaction involving discreet electron transfer from the diene to the catalyst as the initiation step (Scheme 7) (61). Evidence in favor of such a scheme was the direct observation of a carbon-centered free radical (57) ($g = 2.004$) with olefinic hydrogens missing from the proton NMR spectrum and the formation of the same endo peroxide product (58) by anodic oxidation to give this radical cation.

A number of copper-catalyzed oxygenations have been reported which lead to ring cleavage of phenols (62, 63), reactions which appear to be similar to biological phenol cleavages mediated by phenolase and pyrocatechase (64). It has been shown that the treatment of catechols with Cu(I) and oxygen in pyridine–methanol led to *cis,cis*-muconic acid monomethyl ester (59) (65, 66):

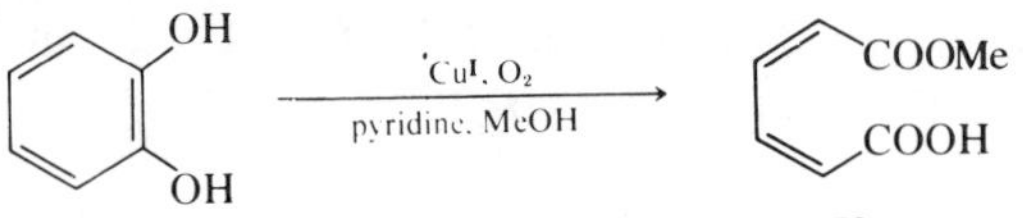

$$\text{catechol} \xrightarrow[\text{pyridine, MeOH}]{Cu^I, O_2} \text{59 (COOMe, COOH diene)}$$

The role of oxygen in this reaction appears to be the reoxidation of Cu(I) to Cu(II), since it has been shown that with stoichiometric amounts of cupric methoxy chloride, ring cleavage is observed even under anaerobic conditions. Phenol was also found to give *cis,cis*-muconic acid under aerobic conditions (67, 68):

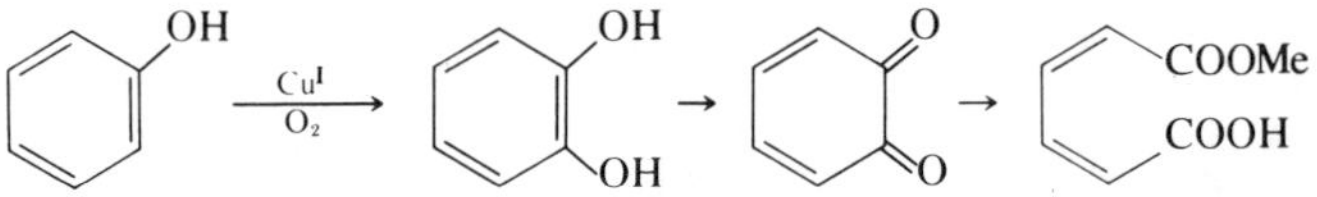

$$\text{PhOH} \xrightarrow{Cu^I, O_2} \text{catechol} \rightarrow \text{o-quinone} \rightarrow \text{COOMe, COOH diene}$$

An important conclusion to be drawn from these results is that no active oxygen species is involved in this ring cleavage reaction, but rather an active metal species.

5 HYDROXYLATIONS WITH FERROUS ION–HYDROGEN PEROXIDE

Several years ago we began a series of experiments designed to probe the nature of reactive intermediates in iron–peroxide systems. It has long been known that the ferrous ion–hydrogen peroxide mixtures known as Fenton's reagent (69, 70) produced a strongly oxidizing medium capable of oxidizing most organic compounds. The generally accepted mechanism

HYDROXYLATIONS WITH FERROUS ION–HYDROGEN PEROXIDE

for Fenton's reagent was that proposed by Haber and Weiss and involved initial reductive cleavage of the peroxy bond by ferrous ion to produce ferric ion, hydroxide ion, and a hydroxyl radical (71):

$$Fe^{2+} + H_2O_2 \rightarrow Fe^{3+} + HO^- + HO^\cdot$$

A more recent examination of this reaction by Walling has arrived convincingly at a similar conclusion (72), at least in acidic, aqueous solution. An interesting mechanistic alternative to the hydroxyl radical path was the mechanism proposed by Bray and Gorin in which heterolysis of the peroxy bond was postulated to lead to a ferryl ion species, FeO^{2+} (73):

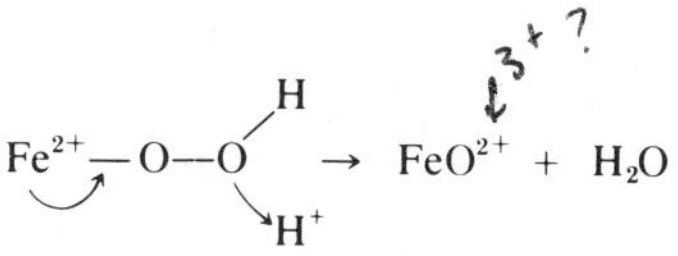

$$Fe^{2+}\!-\!O\!-\!O\!\begin{array}{c}H\\ \\H^+\end{array} \rightarrow FeO^{2+} + H_2O$$

The hydroxy radical is certainly a competent oxidant for all of the transformations carried out by the biological oxygenases, and there is some evidence that hydroxyl radical can be produced by superoxide ion under physiological conditions (74). Indeed there is good reason to suggest that hydroxyl radical production via the Haber–Weiss reaction is an important component in the etiology of oxygen toxicity. It was far from clear how such a process could be mediated by a protein in a selective manner. By contrast, the ferryl ion is ideally suited for directed oxidation reactions since ligation of the iron could provide a means for aligning the oxidant with the substrate. Thus it seemed feasible that an oxidizable ligand could reveal the intermediacy of a ferryl ionlike species if unusual regioselectivity was observed. A number of considerations led us to examine Fenton's reagent oxidations in nonaqueous solution.

The oxidation of cyclohexanol by hydrogen peroxide in the presence of ferrous salts and perchloric acid in acetonitrile was found to give all six possible cyclohexane diols in addition to cyclohexanone (75). The distribution of diol products from this reaction was very unusual, giving 72% *cis*-1,3-cyclohexane diol (**60**), and rather specific for this set of reaction conditions. Substitution of copper for iron, omission of perchloric acid, and substitution of water for acetonitrile all caused a loss of this specificity:

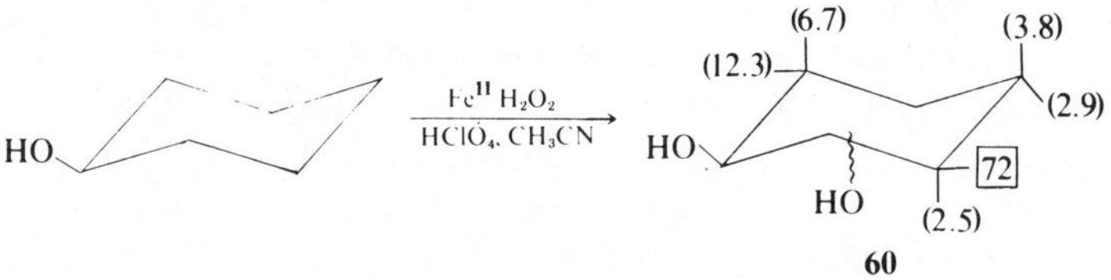

$$\text{HO-cyclohexanol} \xrightarrow{Fe^{II} H_2O_2,\ HClO_4,\ CH_3CN} \text{products}$$

(6.7) (12.3) (3.8) (2.9) (2.5) [72]

60

where the figures in parentheses are the relative reactivities per hydrogen. Deuterium labeling of the starting cyclohexanol established that the product distribution was not the result of rearrangements and that the *cis*-1,3-diol arose from specific removal of the cis hydrogen at carbon 3.

A similar analysis of the hydroxylation of 7-hydroxynorbornane (**61**) by ferrous ion–hydrogen peroxide in acetonitrile led to the conclusion that the syn-exo hydrogens, those nearest to the hydroxyl group, were 5.1 times more reactive than the anti-exo hydrogens on the other side of the molecule as shown in **62**:

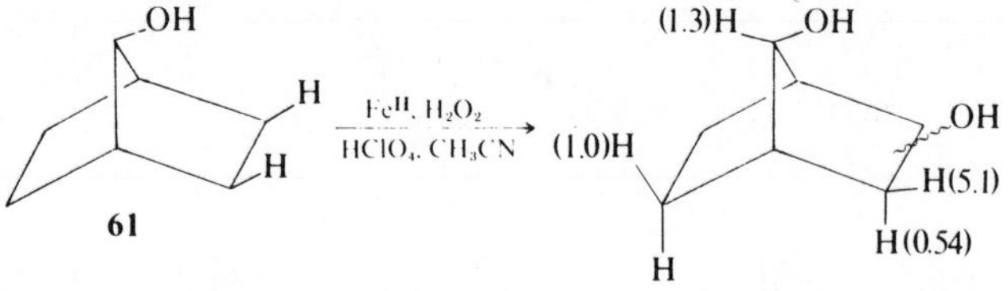

The most reasonable explanation for these regioselectivities is that the hydroxyl group exerts a directive influence on the reaction via ligation of iron (Scheme 8):

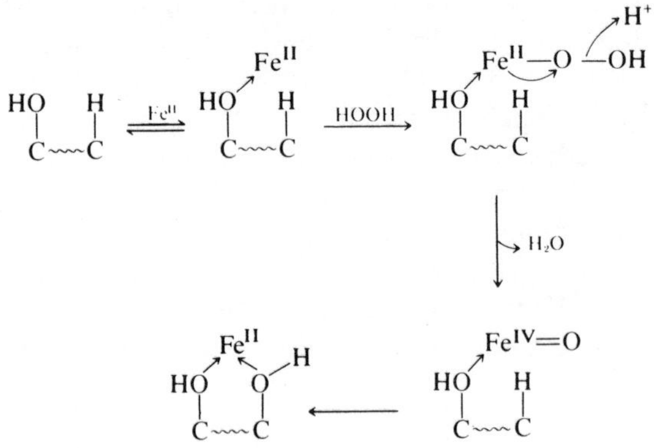

Scheme 8

One can consider two mechanistic extremes for the nature of this directive effect: (a) local generation of hydroxyl radical or (b) two-electron oxidation of the ligated ferrous ion to a reactive Fe(IV) intermediate which was then responsible for oxygen transfer. Several arguments favor the latter possibility. There is ample chemical precedent for the oxidation of Fe(II) and Fe(III) salts to give stable Fe(VI) salts and by inference Fe(IV) or Fe(V) species (76). The electrochemical oxidation of certain iron salts is known to give stable Fe(IV) complexes (77), and the ozonization of Fe(II) has been reported to lead to bridged Fe(III)–O–Fe(III) complexes via an Fe(IV)–oxo intermediate **63** (78):

$$Fe^{II} + O_3 \rightarrow Fe^{IV}=O \xrightarrow{Fe^{II}} Fe^{III}-O-Fe^{III}$$
$$\mathbf{63}$$

Given the indication that ferryl ionlike species could be produced by the hydrogen peroxide oxidation of Fe(II), it was of some interest to determine whether similar species could be produced by independent chemical means. The photoreduction of Fe(III)–OH has been thought to involve the photodissociation of hydroxyl radical, and it seemed in principle that such a photoreduction in acetonitrile might offer an independent route to a hydroxylating iron intermediate. The photoreduction of ferric perchlorate hydrate in the presence of cyclohexanol was found to give a distribution of diol products very similar to that produced by acidic hydrogen peroxide/ferrous ion (79). Significantly, acid was not required for regioselectivity in the photochemical process.

The fact that even Pyrex-filtered light led to rapid photoreduction of Fe(III) under these conditions suggests that the formation of hydroxyl radical in a primary photodissociative step is unlikely. Rather a photochemical disproportionation of an Fe(III) dimer would be energetically more feasible:

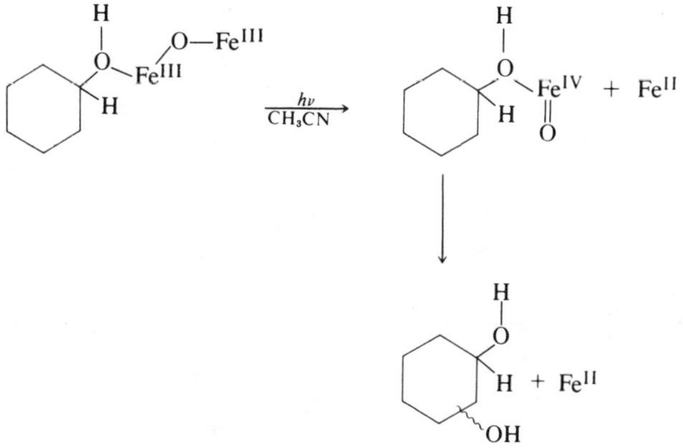

The lack of an acid requirement in the photochemical process is con-

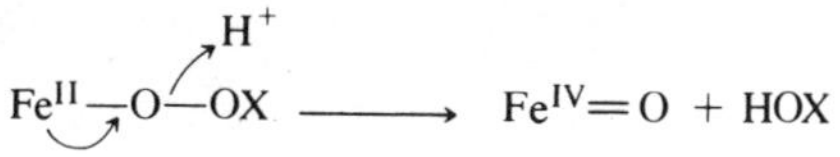

$$Fe^{II}-O-OX \xrightarrow{H^+} Fe^{IV}=O + HOX$$

sistent with the notion of peroxy bond heterolysis in the nonaqueous system. Such a process would be expected to be accelerated by acid catalysis.

6 HYDROXYLATIONS WITH FERROUS IRON–PEROXYACIDS

Support for the idea of peroxy bond heterolysis can be derived from the observation that peroxy acids also gave regioselective hydroxylation of cyclohexanol in the presence of ferrous ion in acetonitrile but with a pronounced preference for oxidation at carbon 2 (80):

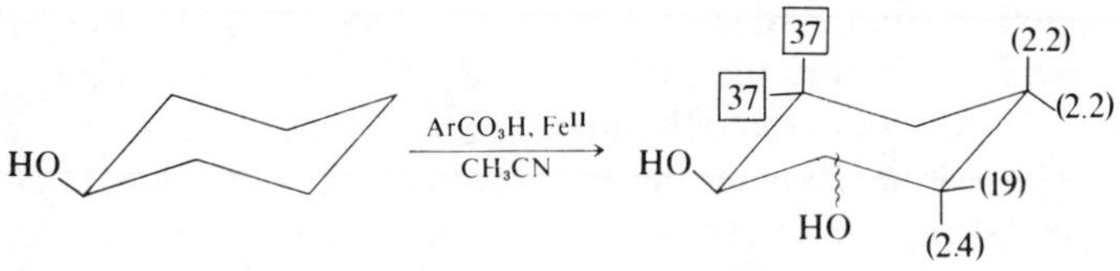

If the hydroxylation of cyclohexanol proceeds as suggested through an

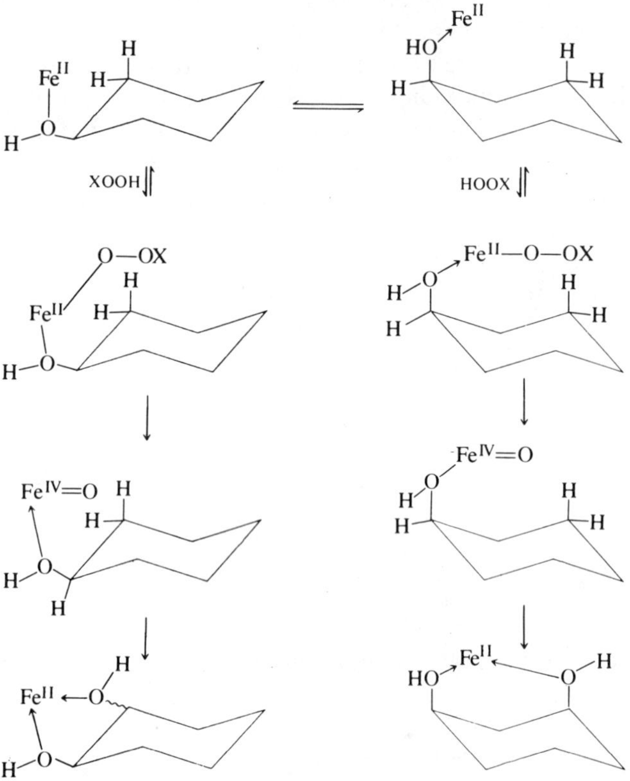

Scheme 9

iron-bound oxidant, two stereoisomeric complexes representing the equatorial and axial conformers of the alcohol must be considered. It can be seen that the distribution of products can be adequately explained by intramolecular oxidation (Scheme 9).

It is of considerable importance to understand the details of oxygen transfer from the intermediate ferryl ion to carbon. Three of the more likely possibilities for this iron–oxo species are (a) a concerted insertion into the C–H bond in a manner reminiscent of the reaction of a singlet carbene (81) [this is the oxenoid insertion mechanism made popular by Hamilton (3) and demonstrated more recently by Olah for hydrogen peroxide in very strong acid media (82)]; (b) a hydride abstraction to give an intermediate carbocation; and (c) a hydrogen atom abstraction to give an intermediate carbon radical:

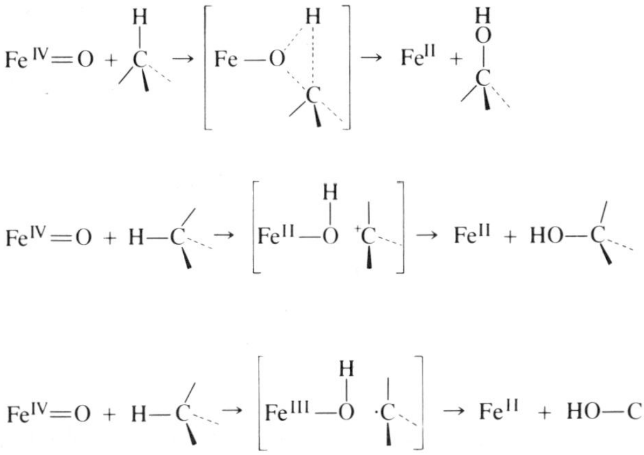

$$Fe^{IV}=O + \overset{H}{\underset{|}{C}} \rightarrow \left[Fe-O \cdots \overset{H}{\underset{|}{C}} \right] \rightarrow Fe^{II} + \overset{H}{\underset{|}{\overset{O}{\underset{|}{C}}}} \quad (a)$$

$$Fe^{IV}=O + H-\overset{}{C} \rightarrow \left[Fe^{II}-\overset{H}{\underset{|}{O}} \; \overset{+}{\underset{|}{C}} \right] \rightarrow Fe^{II} + HO-\overset{}{C} \quad (b)$$

$$Fe^{IV}=O + H-\overset{}{C} \rightarrow \left[Fe^{III}-\overset{H}{\underset{|}{O}} \; \cdot\overset{}{\underset{|}{C}} \right] \rightarrow Fe^{II} + HO-C \quad (c)$$

Several lines of evidence have convinced us that the best description of this aliphatic hydroxylation process is that depicted in path (c). When applied to the peculiar specificity for cis-1,3-diol formation when hydrogen peroxide is the oxidant, the radical path (c) requires that such a radical intermediate be stereoselectively oxidized by Fe(III) irrespective of its mode of generation.

We have found that cis-1,3-cyclohexane diol is produced from either cis- or trans-3-hydroxyperoxycyclohexanecarboxylic acid (**64**) (Scheme 10) (82). Ferric ion is known to be a relatively selective oxidant for carbon radicals (72), and this property serves here to favor a directed oxidation of the incipient 3-hydroxycyclohexyl radical to give predominantly the cis-diol product. Thus it is clear that carbon radicals can be stereoselectively oxidized by Fe(III).

An alternative reaction mechanism, a free-radical chain reaction involving carbon radical displacement on the peroxy acid (84), is less likely

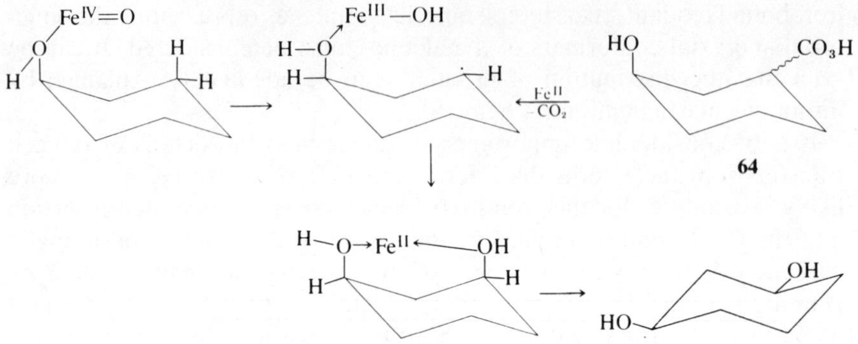

Scheme 10

since the reaction conditions giving the highest yields of *cis*-1,3-diol involved slow addition of the peroxy acid to excess ferrous ion, and under these conditions the peroxy acid reacts instantaneously with ferrous ion:

$$R^{\cdot} + R-C(=O)-O-OH \rightarrow R-OH + R-C(=O)-O^{\cdot}$$

$$R-C(=O)-O^{\cdot} \rightarrow R^{\cdot} + CO_2$$

Evidence against a hydride transfer to give a carbocation is derived from several observations. First, the ratio of 1,2-diol products in the peroxy acid-mediated oxidation is close to unity. The same is true of the reductive decarboxylation of 2-hydroxyperoxycyclohexanecarboxylic acid (82). It is well established that neighboring group participation would strongly favor the *trans*-1,2-diol if a 2-hydroxycyclohexyl carbocation (65) was a discrete intermediate:

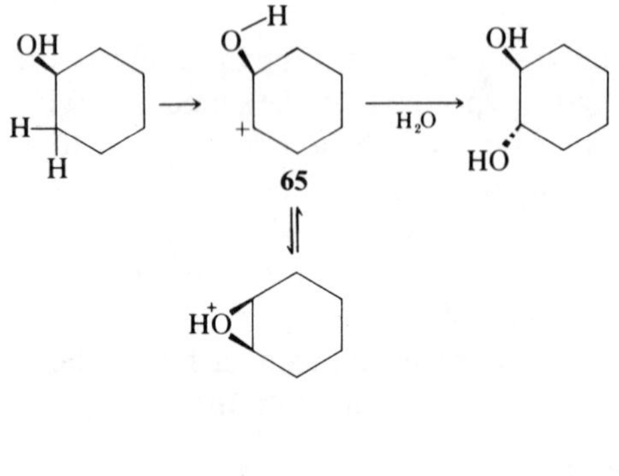

Further, the new hydroxyl group would derive from a nucleophilic oxygen source such as water. We have shown unequivocally that the hydroxylation of cyclohexanol mediated by ferrous ion and peroxy acids involves incorporation of oxygen from the peroxide and only to a small extent from water added to the medium. By contrast, the oxidation of cyclohexene under identical conditions gave only *trans*-1,2-cyclohexane diol with *one* oxygen atom from the peroxide and *one* from the water, as expected for an epoxidation ring opening path (Scheme 11):

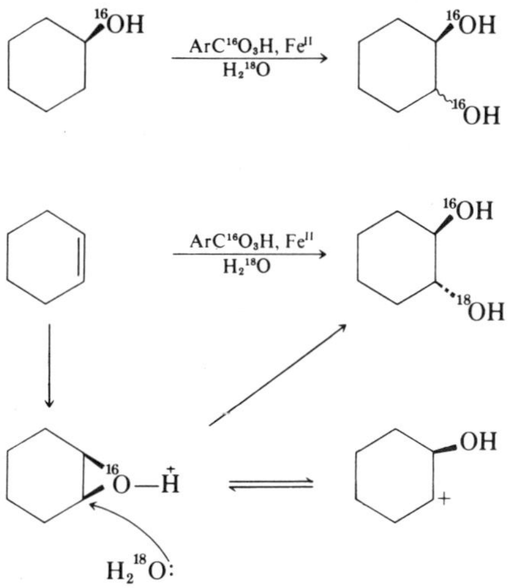

Scheme 11

The significant point here is that the two reactions follow divergent paths and cannot involve a common intermediate. Since the conversion of olefin to diol certainly involves a carbocation intermediate, the aliphatic hydroxylation process cannot also be cationic.

Another indication of the radical nature of this reaction is seen in the oxidation of norcarane (**66**). The copper-catalyzed oxidation of norcarane with *t*-butyl peroxybenzoate is known to give norcaranol (**67**) (85). The mechanism of this process is generally thought to involve a carbon radical intermediate that is oxidized by Cu(II) without passing through a carbocation. The oxidation of norcarane by lead tetraacetate gives 3-cycloheptenol (**68**) via a cationic ring opening (86). We have found that norcarane is hydroxylated by ferrous ion-MCPBA to give only norcaranol (**67**). Thus, as is the case with copper, Fe(III) apparently can oxidize the intermediate norcaranyl radical without producing a cationic intermediate (Scheme 12).

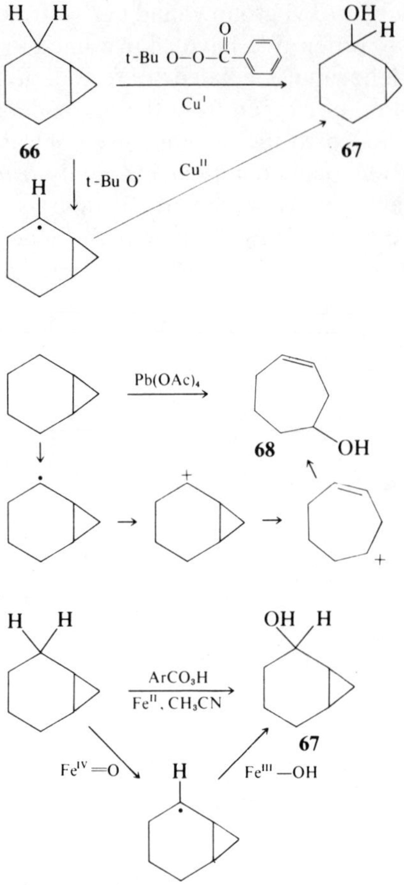

Scheme 12

The oxidation of *trans*-2-deuteriocyclohexanol with *m*-chloroperoxybenzoic acid/ferrous ion and mass-spectral analysis of the diol products revealed that the *cis*-1,2-cyclohexanediol was formed with predominant retention of the deuterium label whereas the *trans*-1,2-cyclohexanediol was formed with significant loss of the label. Thus what appeared superficially to be a nonstereoselective process giving a mixture of stereoisomeric diols was actually two competing processes, each of which proceeded with predominant but not exclusive retention of configuration at the oxidized carbon.

The mechanism we have proposed for this process is outlined in Scheme 13. Intramolecular hydrogen atom abstraction by **69** can give two stereoisomeric radical cage structures, **70** and **71**. These radicals must be very

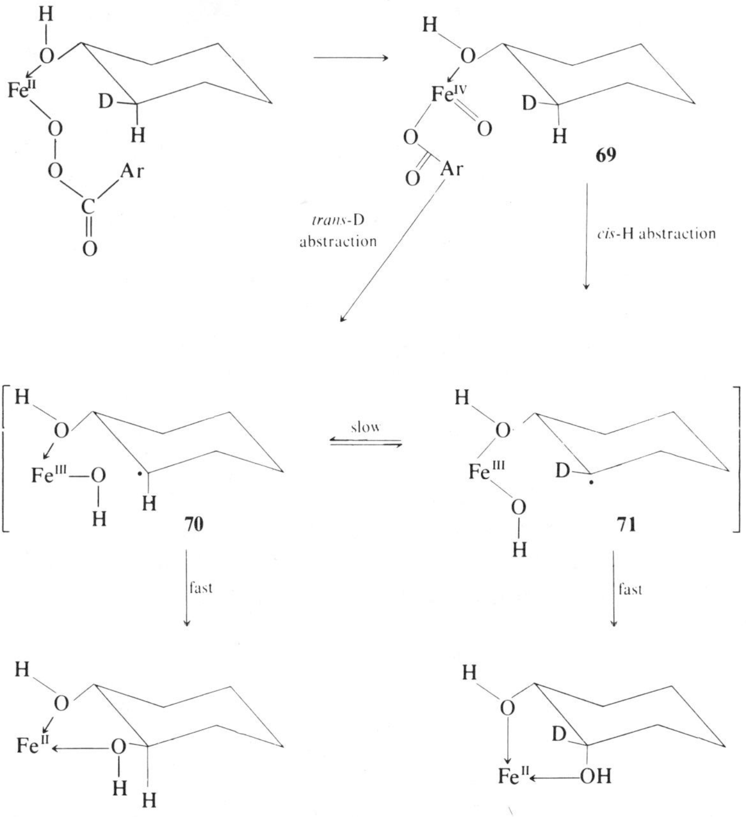

OXIDATIONS CATALYZED BY METALLOPORPHYRINS 155

Scheme 13

short-lived since collapse of the radical cage to give diol products must occur more rapidly than epimerization of the radical in the cage. This result is reminiscent of the alkane hydroxylations by chromic acid (22) and permanganate (23) which were discussed above and further emphasizes analogies between the transient iron–oxo species generated by peroxide oxidation of ferrous ion and the well-documented chemistry of other transition metal–oxo compounds.

7 OXIDATIONS CATALYZED BY METALLOPORPHYRINS

There has been intense interest in oxygen binding and activation by metalloporphyrins since numerous heme-containing enzymes are known to mediate a wide range of oxidative reactions (87). With the clear indication that higher-valent iron species can effect aliphatic hydroxylation in simple systems, it was of interest to know whether similar processes could be catalyzed by iron–porphine complexes. There have been several mentions of the use of metalloporphyrins as hydroxylation catalysts.

Belova has reported the oxidation of cyclohexane with molecular oxygen in the presence of thiosalisylic acid and Fe(III) protoporphyrin IX [Fe^{III}PPIX-Cl] (88). Cyclohexanol and cyclohexanone are the only reported products. The authors state that the reaction was inhibited by cyanide, that radical inhibitors were only moderately effective, and that oxygen could not be replaced by hydrogen peroxide. It was proposed that the metalloporphyrin was serving here as an oxene transfer agent.

More recent studies have shown that iron and cobalt porphyrins are potent autoxidation initiators. Thus with cyclohexene, cyclohexanone, cyclohexanol, and to a lesser extent cyclohexene oxide were the products, and the reaction showed an induction period typical of a free-radical autoxidation (89, 90). In the case of chlorotetraphenylporphinato iron(III) [Fe^{III}TPP-Cl], the starting monomeric catalyst was converted to the μ-oxo dimer during this induction period. Cyclohexenyl hydroperoxide was shown to be the first-formed product in these reactions:

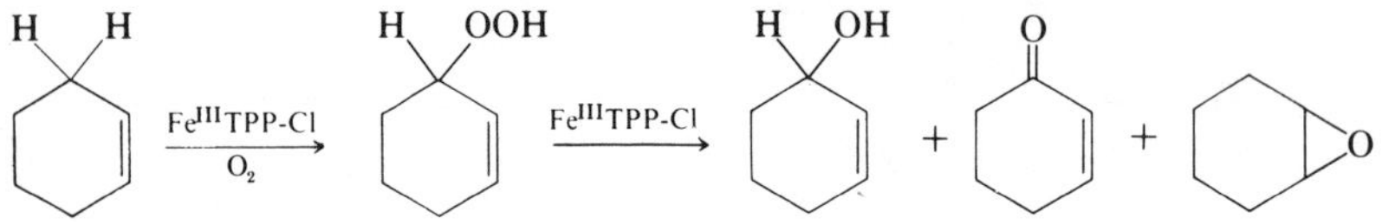

In an attempt to produce an iron–porphine–oxo species from typical porphyrins like Fe^{III}TPP-Cl and Fe^{III}PPIX-Cl, we have examined the reaction of t-butyl hydroperoxide and peroxy acids with alkanes and olefins. With peroxy acids, decomposition of the porphyrin ring was observed, while with t-butyl hydroperoxide, product distributions were indistinguishable from free-radical chain reactions initiated photochemically in the absence of any metals.

$$t\text{-Bu OOH} \longrightarrow t\text{-Bu O}^{\cdot}$$

$$t\text{-Bu O}^{\cdot} + R\text{—H} \longrightarrow R^{\cdot} + t\text{-Bu OH}$$

$$R^{\cdot} + t\text{-Bu OOH} \longrightarrow R\text{—OH} + t\text{-Bu O}^{\cdot}$$

It appears therefore that the redox properties of the metalloporphyrin are required only for the initiation step in these free-radical autoxidations and that the porphyrin is not a stoichiometrically significant catalyst.

The failure of these simple approaches to a reactive iron–porphine oxide or its equivalent is intriguing in light of several reports that cytochrome P-450 can hydroxylate substrates anaerobically in the presence of such oxygen donors (91, 92). Indeed, it could easily be that axial ligation, presumably by thiolate in cytochrome P-450 (93, 94), is crucial to oxygen transfer by this enzyme or that other functionality on nearby protein plays an important catalytic role.

We have recently demonstrated however that epoxidation and hydroxylation can be achieved with simple iron–porphine catalysts with iodosylbenzene as the oxidant (95). Thus for example, reaction of cyclohexene with iodosylbenzene in the presence of catalytic amounts of Fe^{III}TPP-Cl gave cyclohexene oxide and cyclohexenol in 55 and 15% yields, respectively. Iodobenzene was recovered quantitatively. Likewise, cyclohexane is converted to cyclohexanol under these conditions. Significantly, the alcohols were not rapidly oxidized to ketones under these conditions, a selectivity shared with the enzymic hydroxylations:

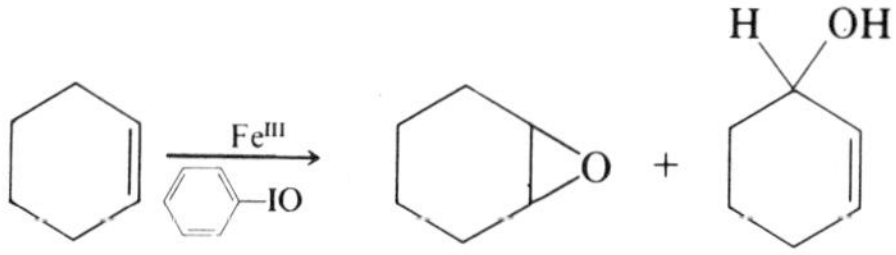

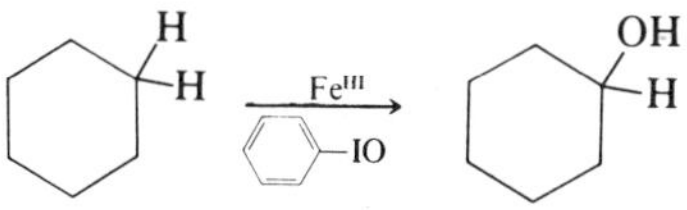

Remarkably *cis*-double bonds were found to be more reactive than *trans*-double bonds with Fe^{III}TPP-Cl but not with Fe^{III}PPIX-Cl. A particularly spectacular example was the difference in reactivity between *cis*- and *trans*-stilbene. *cis*-Stilbene was epoxidized to give *cis*-stilbene oxide in yields greater than 80% with Fe^{III}TPP-Cl/iodosylbenzene. In a competitive oxidation of these two stereoisomers, only trace amounts of *trans*-stilbene oxide were produced, and the *trans*-stilbene was recovered unreacted:

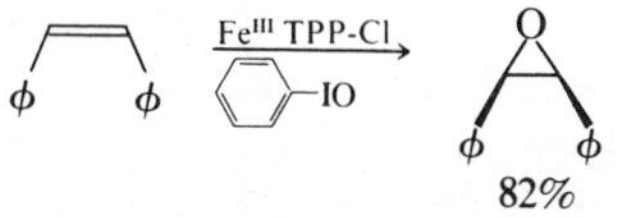

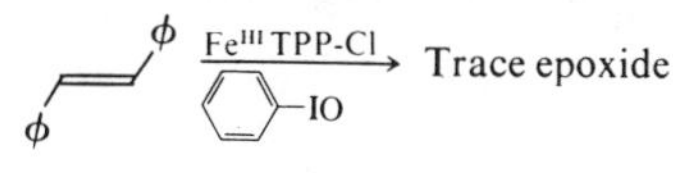

The fact that Fe^{III}-PPIX-Cl showed no such specificity indicates that these epoxidations are sensitive to the substitution pattern on the porphyrin periphery and that the oxygen transfer must be occurring at the metalloporphyrin center.

As a test of this hypothesis, we have oxidized dioctyl Fe^{III}PPIX-Cl with iodosylbenzene. Analysis of the products showed that the octyl side chains had been hydroxylated and that 60% of the hydroxylation had occurred at carbon 4 and carbon 5 in the middle of the chain! Examination of molecular models indicates that these two carbon centers have the easiest access to the center of the porphyrin, a convincing indication that the mechanism of this hydroxylation is an intramolecular oxygen rebound (80) from iodine to iron and into the C–H bond (Scheme 14):

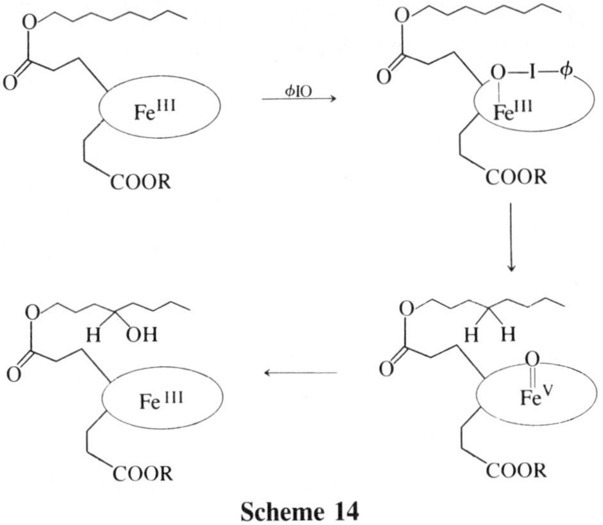

Scheme 14

It is interesting to note that while iodosylbenzene is a rather unreactive oxidant (96), it reacts rapidly with iron porphyrins to give an oxidant capable of hydroxylating unactivated carbon–hydrogen bonds. The most reasonable interpretation is that iodosylbenzene forms a complex with the porphyrin and that this complex decomposes to an Fe(V)–oxo intermediate which is intrinsically reactive toward hydrocarbons. Indeed we have recently obtained spectral evidence for the formation of a reactive intermediate in this reaction (97, 98). The detailed nature of oxygen transfer catalyzed by porphyrins is under current investigation in our laboratories.

ACKNOWLEDGMENTS

The portions of the work described herein that were carried out in the author's laboratories were supported by the National Science Foundation (CHE 77-21849), the National Institutes of Health (1R01-GM 25923-01), and the Petroleum Research Fund administered by the American Chemical Society.

REFERENCES

1. O. Hayaishi, in *Oxygenases*, O. Hayaishi, Ed., Academic Press, New York–London, 1976, pp. 1–29.
2. R. Vercauteren and L. Massart, in *Oxygenases*, O. Hayaishi, Ed., Academic Press, New York–London, 1962, pp. 355–407.
3. G. A. Hamilton, in *Molecular Mechanisms of Oxygen Activation*, O. Hayaishi, Ed., Academic Press, New York–London, 1974, pp. 405–451.
4. A. B. P. Lever and H. B. Gray, *Acc. Chem. Res.*, **11**, 348 (1978).
5. L. Vaska, *Acc. Chem. Res.*, **9**, 175 (1976).
6. G. Henrici-Olive and S. Olive, *Angew. Chem.*, **86**, 1 (1974).
7. S. L. Regen and G. M. Whitesides, *J. Organometal. Chem.*, **59**, 293 (1973).
8. R. Sheldon and J. A. Von Doorn, *J. Organometal. Chem.*, **94**, 115 (1975).
9. R. Ugo, *Englehard Industries Technical Bulletin*, **XI**(2), 45 (1971) (through reference 8).
10. C. E. Griffin and S. K. Kundu, *J. Org. Chem.*, **34**, 1532 (1969).
11. H. C. Stevens and A. J. Kamau, *J. Am. Chem. Soc.*, **87**, 734 (1965).
12. H. Mimoun, I. Seree deRoch, and L. Sajus, *Tetrahedron*, **26**, 37 (1970).
13. K. B. Sharpless, J. M. Townsend, and D. R. Williams, *J. Am. Chem. Soc.*, **94**, 295 (1972).
14. A. K. Awasthy and J. Roček, *J. Am. Chem. Soc.*, **91**, 991 (1969).
15. A. O. Chong and K. B. Sharpless, *J. Org. Chem.*, **42**, 1587 (1977).
16. G. Read and P. J. C. Walcker, *J. Chem. Soc. Dalton Trans.*, 883 (1977).
17. D. Holland and D. J. Milner, *J. Chem. Soc. Dalton Trans.*, 2440 (1975); J. Farrar, D. Holland, and D. J. Milner, *ibid.*, 815 (1975).
18. H. Mimoun, M. M. P. Machirant, and I. Seree deRoch, *J. Am. Chem. Soc.*, **100**, 5437 (1978).
19. F. Ingersheim and H. Mimoun, *J. Chem. Soc. Chem. Commun.*, 559 (1978).
20. G. A. Olah and J. Welch, *J. Org. Chem.*, **43**, 2830 (1978).
21. M. J. Y. Chen and J. K. Kochi, *J. Chem. Soc. Chem. Commun.*, 204 (1977).
22. K. B. Wiberg, in *Oxidation in Organic Chemistry, Part A*, K. B. Wiberg, Ed., Academic Press, New York–London, 1965, pp. 69–184.
23. R. Stewart, in *Oxidation in Organic Chemistry, Part A*, K. B. Wiberg, Ed., Academic Press, New York–London, 1965, pp. 2–68.

24. F. Mares and J. Roček, *Collect. Czech. Chem. Commun.*, **26**, 2370 (1961).
25. K. B. Wiberg and G. Foster, *J. Am. Chem. Soc.*, **83**, 423 (1961).
26. F. Mares, J. Roček, and J. Sicher, *Collect. Czech. Chem. Commun.*, **26**, 2355 (1961).
27. K. B. Wiberg and G. Szeimies, *J. Am. Chem. Soc.*, **96**, 1889 (1974).
28. J. K. Kochi, in *Free Radicals*, Vol. I, J. K. Kochi, Ed., Wiley, New York, 1973, pp. 591–683.
29. J. K. Kochi and J. D. Bacha, *J. Org. Chem.*, **33**, 2246 (1968).
30. C. L. Jenkins and J. K. Kochi, *J. Am. Chem. Soc.*, **94**, 856 (1972).
31. J. K. Kochi and D. Mog, *J. Am. Chem. Soc.*, **87**, 522 (1965).
32. K. B. Sharpless, A. Y. Teranishi, and J. E. Bäckvall, *J. Am. Chem. Soc.*, **99**, 3120 (1977).
33. W. A. Waters and J. S. Littler, in *Oxidation in Organic Chemistry, Part A*, Academic Press, New York–London, 1965, pp. 185–241.
34. J. I. Brauman and A. J. Pandell, *J. Am. Chem. Soc.*, **92**, 329 (1970).
35. R. J. Audette, J. W. Quail, and P. J. Smith, *Tetrahedron Lett.*, 279 (1971).
36. R. J. Audette, J. W. Quail, and P. J. Smith, *Chem. Commun.*, 38 (1972).
37. J. Riley, Ph.D. Thesis, University of Kentucky, 1968.
38. K. B. Sharpless and T. C. Flood, *J. Am. Chem. Soc.*, **93**, 231 (1971).
39. D. M. Stone, unpublished results.
40. J. Yoder, unpublished results.
41. S. Udenfriend, C. T. Clark, J. Axelrod, and B. B. Brodie, *J. Biol. Chem.*, **208**, 731 (1954).
42. G. A. Hamilton, R. J. Workman, and L. Woo, *J. Am. Chem. Soc.*, **86**, 3390 (1964).
43. G. A. Hamilton, J. P. Friedman, and P. M. Campbell, *J. Am. Chem. Soc.*, **88**, 5266 (1966).
44. G. A. Hamilton, J. W. Hanifin, Jr., and J. P. Friedman, *J. Am. Chem. Soc.*, **88**, 5269 (1966).
45. U. Frommer and V. Ullrich, *Z. Naturforsch.*, **26b**, 322 (1971).
46. M. Viscortini, *Angew. Chem.*, **80**, 492 (1968).
47. M. B. Dearden, C. R. E. Jefcoate, and J. R. Lindsay Smith, *Adv. Chem. Ser.*, **77**, III, 260 (1968).
48. H. Mimoun and I. Seree deRoch, *Tetrahedron*, **31**, 377 (1975).
49. T. Matsuura, *Tetrahedron*, **33**, 2869 (1977).
50. C. Walling, G. M. El-Taliawi, and R. A. Johnson, *J. Am. Chem. Soc.*, **33** (1974).
51. J. R. Lindsay Smith, B. A. J. Shaw, A. M. Jeffrey, and D. M. Jerina, *J. Chem. Soc. Perkin II*, 1583 (1977).
52. A. Nishinaga, K. Nishigawa, H. Tomita, and T. Matsuura, *J. Am. Chem. Soc.*, **99**, 1287 (1977).
53. A. Nishinaga, *Chem. Lett.*, 273 (1975).
54. F. Hirata, T. Ohnishi, and O. Hayaishi, *J. Biol. Chem.*, **252**, 4639 (1977).
55. T. Ohnishi, F. Hirata, and O. Hayaishi, *J. Biol. Chem.*, **252**, 4643 (1977).
56. O. Hayaishi, F. Hirata, T. Ohnishi, J. P. Henry, I. U. Rosenthal, and A. Kath, *J. Biol. Chem.*, **252**, 3548 (1977).

REFERENCES

57. T. Morimoto and Y. Ogata, *J. Chem. Soc. (B)*, 62 (1967).
58. D. H. R. Barton, R. K. Haynes, P. D. Magnus, and I. D. Menzies, *J. Chem. Soc. Chem. Commun.*, 511 (1974).
59. D. H. R. Barton, R. K. Kaynes, G. Leclerc, P. D. Magnus, and I. D. Menzies, *J. Chem. Soc. Perkin Trans. I*, 2055 (1975).
60. R. K. Haynes, *Aust. J. Chem.*, **31**, 121, 131 (1978).
61. R. Tang, H. J. Yue, J. F. Wolf, and F. Mares, *J. Am. Chem. Soc.*, **100**, 5248 (1978).
62. R. R. Grinstead, *Biochemistry*, **3**, 1308 (1964).
63. J. Tsuji, H. Takayanagi, and I. Sakai, *Tetrahedron Lett.*, 1245 (1975).
64. O. Hayaishi, M. Katagori, and S. Rothberg, *J. Am. Chem. Soc.*, **77**, 5450 (1955).
65. J. Tsuji and H. Takayanagi, *J. Am. Chem. Soc.*, **96**, 7349 (1974).
66. J. Tsuji and H. Takayanagi, *Tetrahedron Lett.*, 1365 (1976).
67. M. M. Rogić, T. R. Demmin, and W. B. Hammond, *J. Am. Chem. Soc.*, **98**, 7441 (1976).
68. M. M. Rogić and T. R. Demmin, *J. Am. Chem. Soc.*, **100**, 5472 (1978).
69. H. J. H. Fenton, *J. Chem. Soc.*, **65**, 899 (1894).
70. H. J. H. Fenton, *J. Chem. Soc.*, **69**, 546 (1896).
71. F. Haber and J. Weiss, *Proc. Roy. Soc.* (London), **A147**, 332 (1934).
72. C. Walling, *Acc. Chem. Res.*, **8**, 125 (1975).
73. W. C. Bray and M. H. Gorin, *J. Am. Chem. Soc.*, **54**, 2124 (1932).
74. B. Halliwell, *FEBS Lett.*, **92**, 321 (1978).
75. J. T. Groves and M. Van Der Puy, *J. Am. Chem. Soc.*, **96**, 5274 (1974); *ibid.*, **98**, 5290 (1976).
76. G. L. Kochanny and A. Timnick, *J. Am. Chem. Soc.*, **83**, 2777 (1961).
77. L. F. Warren and M. A. Bennett, *J. Am. Chem. Soc.*, **96**, 3340 (1974).
78. T. J. Conocchioli, E. S. Hamilton, and N. Sutin, *J. Am. Chem. Soc.*, **87**, 926 (1965).
79. J. T. Groves and W. W. Swanson, *Tetrahedron Lett.*, 1953 (1975).
80. J. T. Groves and G. A. McClusky, *J. Am. Chem. Soc.*, **98**, 859 (1976).
81. C. D. Gutsche, G. L. Beeman, W. Udell, and S. Baurlein, *J. Am. Chem. Soc.*, **93**, 5172 (1971).
82. G. Olah, N. Yoneda, and D. G. Parker, *J. Am. Chem. Soc.*, **99**, 483 (1977).
83. J. T. Groves and M. Van Der Puy, *J. Am. Chem. Soc.*, **97**, 7118 (1975).
84. T. M. Luong and D. Lefort, *Bull. Soc. Chim. Fr.*, 827 (1962).
85. T. Shono, Y. Matsumura, and Y. Nakagawa, *J. Org. Chem.*, **36**, 1771 (1971).
86. R. J. Ouellette, A. South, and D. L. Shaw, *J. Am. Chem. Soc.*, **87**, 2602 (1965).
87. J. T. Groves, in *Advances in Inorganic Biochemistry*, Vol. I, G. L. Eichhorn and L. G. Marzilli, Eds., Elsevier, North-Holland, New York, 1979, pp. 119–145.
88. V. S. Belova, L. A. Nikonova, L. M. Raikhman, and M. R. Borukaeva, *Dokl. Aked. Nauk SSR (Engl. Ed.)*, **204**, 455 (1972).
89. D. R. Paulson, R. Ullman, R. B. Sloane, and G. L. Closs, *J. Chem. Soc. Chem. Comm.*, 186 (1974).
90. Y. Ohkatsu and T. Tsuruta, *Bull. Chem. Soc. Japan*, **51**, 188 (1978).
91. E. G. Hrycay and P. J. O'Brien, *Arch. Biochem. Biophys.*, **153**, 480 (1972).

92. F. Lichtenberger, W. Nastainczyk, and V. Ullrich, *Biochem. Biophys. Res. Commun.*, **70**, 939 (1976).
93. K. Murikami and H. S. Mason, *J. Biol. Chem.*, **242**, 1102 (1967).
94. P. M. Champion, I. C. Gunzalus, and G. C. Wagner, *J. Am. Chem. Soc.*, **100**, 3743 (1978), and references therein.
95. J. T. Groves, T. E. Nemo, and R. S. Myers, *J. Am. Chem. Soc.*, **101**, 1032 (1979).
96. T. Takaya, H. Enyo, and E. Imoto, *Bull. Chem. Soc. Japan*, **41**, 1032 (1968).
97. T. E. Nemo, unpublished results.
98. J. T. Groves and W. J. Krupes, Jr., *J. Am. Chem. Soc.*, **101**, 7613 (1979).

CHAPTER 4
Recent Progress on the Mechanism of Action of Dioxygenases

J. M. WOOD

Gray Freshwater Biological Institute, Department of Biochemistry,
University of Minnesota, Navarre, Minnesota

CONTENTS

1. OXYGENASES, 165
2. "ACTIVATION" OF OXYGEN, 166
3. DIOXYGENASE STRUCTURE, 167

    3.1 PCA 3,4-Oxygenase, 167
    3.2 PCA 4,5-Oxygenase, 170
    3.3 PCA 2,3-Oxygenase, 171

4. SUBSTRATE BINDING AND KINETICS, 171
5. DIOXYGENASE MECHANISMS, 175

    ACKNOWLEDGMENTS, 178

    REFERENCES, 178

1 OXYGENASES

Enzymes that catalyze the insertion of molecular oxygen into organic substrates are called "oxygenases." These enzymes provide examples for the historical concept of "oxygen fixation" first developed by Lavoisier, who defined oxidation as the addition of oxygen to a chemical substance.

In the 1950s Osamu Hayaishi's group in Kyoto discovered the enzyme pyrocatechase, which was shown to catalyze the cleavage of catechol with the direct incorporation of dioxygen to give *cis,cis*-muconic acid as the ring fission product (1). Later the same research team used $^{18}O_2$ to show that both atoms of the same oxygen molecule were incorporated directly into the reaction product (2) (Scheme 1):

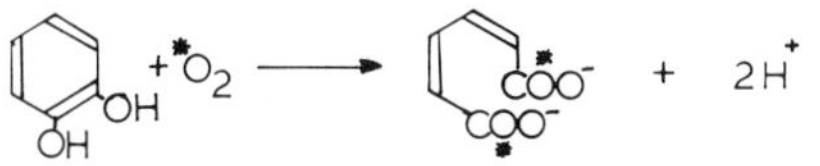

Scheme 1. $^{18}O_2$ insertion by the "intradiol" mechanism.

Enzymes that directly fix molecular oxygen into organic substrates have been found to be ubiquitous, playing a crucial role in the degradation of natural products in the biosphere (3). Aerobic microorganisms have the ability to degrade stable organic compounds such as lignin, alkaloids, flavanoids, xanthones, and terpenes. These degradative reactions are necessary to maintain the balance of the carbon cycle (4). In addition dioxygenases function in the oxidation of thiols and in the hydroperoxidation of polyunsaturated fatty acids.

Although the central position of oxygenases is to catalyze the degradation of natural products, in advanced industrial societies these enzymes are also required to accommodate the oxidation of industrial chemicals and pollutants (5). Pollutants that are toxic to microorganisms are often detoxified by hydroxylation to products that are more soluble in water. For example the oxidation of toluene to the methylcatechols yields water-soluble products which become diluted in the aqueous environment (6). Very often dihydroxylated as well as monohydroxylated products are substrates for dioxygenases. Recently it has been discovered that the chloro phenol, 5-chlorosalycyclic acid, is cleaved by a specific dioxygenase to yield a ring fission product which eliminates chlorine as chloride ion through lactonization and hydrolysis (7) (Scheme 2):

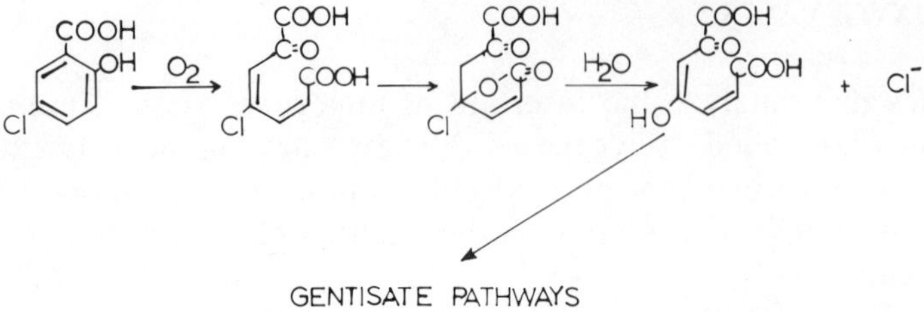

Scheme 2 Catabolism of 5-chlorosalicylate by extracts of *Bacillus brevis*.

Therefore, stable chlorinated aromatic compounds of industrial origin can be dehalogenated by conversion to aliphatic products which enter central metabolic pathways.

A better understanding of the mechanisms of action of the oxygenases could help us to predict the biodegradability of man-made materials. Also this technology could be applied in the selective oxidation of natural products such as lignin. The removal of lignin from woody plants would allow more efficient utilization of celluloses and hemicelluloses as sources of food, energy, and paper (8). However we shall only make progress in this important field when we understand how molecular oxygen is "activated" and inserted into stable organic compounds. To obtain this knowledge a detailed understanding of the structure and function of the oxygenases is required.

2 "ACTIVATION" OF OXYGEN

The fact that organic compounds are relatively stable in the biosphere indicates that the reaction of these compounds with molecular oxygen proceeds by reactions which are generally unfavorable from a kinetic standpoint. This is not too difficult to rationalize, because dioxygen exists as a triplet in the ground state whereas carbon compounds exist in the singlet state. This makes concerted reactions between oxygen and organic carbon compounds difficult without some kind of "activation." In the enzyme-catalyzed reactions this difficulty is circumvented either by reacting dioxygen with the unpaired d electrons of a transition metal ion complex, or by reacting dioxygen with a stable organic free radical. Organic free radicals are best stabilized in highly conjugated systems (i.e., coenzymes such as FMN or FAD). In each case we are considering free-radical reactions in the enzymes. The active sites of such enzymes are designed to channel the free-radical reaction to yield a specific product, and not the mixture of products which would most likely be produced if reactions of this kind were attempted with transition metal complexes or free radicals in solution.

3 DIOXYGENASE STRUCTURE

During the last three years a careful study of the structure of three dioxygenases which cleave protocatechuic acid (PCA) has been carried out (8). The alternate ring cleavage positions for these three enzymes are presented in Scheme 3:

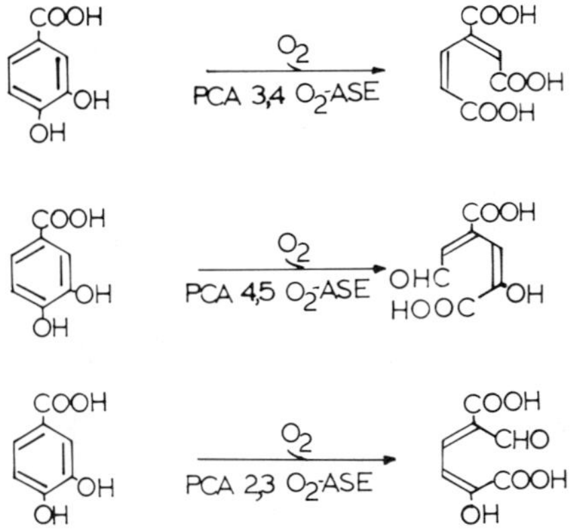

Scheme 3 Dioxygenase reactions catalyzed by PCA3,4-, 4,5-, and 2,3–oxygenases.

In the following pages we examine the protein chemistry of PCA 3,4-, 4,5-, and 2,3-oxygenases.

3.1 PCA 3,4-Oxygenase

PCA 3,4-oxygenase was first discovered by Stanier and Ingraham when it was partially purified from cell extracts of *Pseudomonas fluorescens* grown with *p*-hydroxybenzoate as sole carbon and energy source (9). Fujisawa and Hayaishi purified and crystallized the enzyme from *Pseudomonas aeruginosa* in 1968 and showed that it had a molecular weight of approximately 700,000. The crystalline enzyme is deep red in color with a visible absorption band from 400 to 650 nm and a λ_{max} at 450 nm. The native enzyme was shown to contain approximately 7 g-atoms of iron per enzyme molecule (10). Total amino acid analysis gave 95 cysteine residues per molecule, 12 of which could be titrated with *p*-chloromercuribenzoate; after treatment with 8 *M* urea an additional 43 residues were titrated, but the remaining 40 residues were only available for titration after reduction with sodium borohydride. This experiment implied the presence of a number of crosslinking disulfide bridges. The presence of 12 free sulfhydryl groups provides an explanation for the formation of polymeric forms of the enzyme which precipitate when mercaptoethanol is omitted from crystallization procedures (10).

In 1976 Nozaki et al. (11) and Lipscomb et al. (12) independently discovered that PCA 3,4-oxygenase was composed of nonidentical subunits of molecular weights 22,500 and 25,000. These subunits were designated α and β. Lipscomb (12) showed that the native enzyme dissociated into an $\alpha_2\beta_2$ structure. Recently Kohlmiller and Howard (13) have sequenced the α subunit, and Lorsbach and Howard have completed 70% of the sequence of the β subunit. The primary structure of the α subunit is presented in Scheme 4.

Rao and Sundaralingham have examined crystalline PCA 3,4-oxygenase and have shown that this enzyme may be amenable to X-ray structural analysis in the future (14). A general picture of the basic structure of PCA 3,4-oxygenase can now be postulated. Each $\alpha_2\beta_2$ structure contains one active iron center in the native enzyme and eight $\alpha_2\beta_2$ structures associate in the formation of the crystalline protein.

Ballou and Bull (15) have isolated PCA 3,4-oxygenase from *Pseudomonas putida*. This enzyme preparation has a molecular weight of 190,000 and is composed of two $\alpha_2\beta_2$ structures compared with eight $\alpha_2\beta_2$ structures found in *Pseudomonas aeruginosa*. It is interesting to note that the enzyme from *Ps. putida* contains 2 g-atoms iron per $\alpha_2\beta_2$ structure, whereas the enzyme from *Ps. aeruginosa* contains only 1 g-atom per tetramer. The native enzyme from *Ps. aeruginosa* will take up an additional g-atom iron per $\alpha_2\beta_2$ unit, but the specific activity of the enzyme is only increased 25% upon reconstitution with this extra iron.

The similarity in the optical spectrum of PCA 3,4-oxygenase to that of the various transferrins indicated that tyrosine may be coordinated to the active iron center (8). Certainly the optical spectrum of the native enzyme could be explained as a tyrosine $\rightarrow$ Fe(III) charge transfer band. Keyes et al. (16), Ballou and Bull (15), and Que and Heistand (17) have shown that tyrosine is coordinated to the high-spin Fe(III) centers of the intradiol-cleaving enzymes PCA 3,4-oxygenase (*Ps. aeruginosa*), PCA 3,4-oxygenase (*Ps. putida*), and pyrocatechase (*Ps. arvilla* C_1). Resonance Raman was used to give enhanced peaks around 1602, 1503, 1263, and 1171 cm^{-1} upon laser excitation at 514.5 nm. These frequencies are characteristic of tyrosinate $\rightarrow$ Fe(III) coordination in a number of transferrins (18–20). Clearly we have a new class of nonheme iron proteins with tyrosinate $\rightarrow$ Fe(III) coordination. Table 1 shows a comparison of the prin-

PROTOCATECHUIC ACID 3,4 OXYGENASE (PRIMARY SEQUENCE OF THE α SUBUNIT)

PRO-	ILE-	GLU-	LEU-	LEU-	PRO-	GLU-	THR-	PRO-	SER-	10
GLN-	THR-	ALA-	GLY-	PRO-	TYR-	VAL-	HIS-	ILE-	GLY-	20
LEU-	ALA-	LEU-	GLU-	ALA-	ALA-	GLY-	ASN-	PRO-	THR-	30
ARG-	ASP-	GLN-	GLU-	ILE-	TRP-	ASN-	ARG-	LEU-	ALA-	40
LYS-	PRO-	ASP-	ALA-	PRO-	GLY-	GLU-	HIS-	ILE-	LEU-	50
LEU-	LEU-	GLY-	GLN-	VAL-	TYR-	ASP-	GLY-	ASX-	HIS-	60
GLY-	LEU-	VAL-	ARG-	ASP-	SER-	PHE-	LEU-	GLU-	VAL-	70
TRP-	GLN-	ALA-	ASP-	ALA-	ASP-	GLY-	GLU-	TYR-	GLN-	80
ASP-	ALA-	TYR-	ASN-	LEU-	GLU-	ASN-	ALA-	PHE-	ASN-	90
SER-	PHE-	GLY-	ARG-	THR-	ALA-	THR-	THR-	PHE-	ASP-	100
ALA-	GLY-	GLU-	TRP-	THR-	LEU-	HIS-	THR-	VAL-	LYS-	110
PRO-	GLY-	VAL-	VAL-	ASN-	ASN-	ALA-	ALA-	GLY-	VAL-	120
PRO-	MET-	ALA-	PRO-	HIS-	ILE-	ASN-	ILE-	SER-	LEU-	130
PHE-	ALA-	ARG-	GLY-	ILE-	ASN-	ILE-	HIS-	LEU-	HIS-	140
THR-	ARG-	LEU-	TYR-	PHE-	ASP-	ASP-	GLU-	ALA-	GLN-	150
ALA-	ASN-	ALA-	LYS-	CYS-	PRO-	VAL-	LEU-	ASN-	LEU-	160
ILE-	GLU-	GLN-	PRO-	GLN-	ARG-	ARG-	GLU-	THR-	LEU-	170
ILE-	ALA-	LYS-	ARG-	CYS-	GLU-	VAL-	ASP-	GLY-	LYS-	180
THR-	ALA-	TYR-	ARG-	PHE-	ASP-	ILE-	ARG-	ILE-	GLN-	190
GLY-	GLU-	GLY-	GLU-	THR-	VAL-	PHE-	PHE-	ASP-	PHE	200

Scheme 4 The primary sequence of the α subunit of PCA 3,4-oxygenase from *Pseudomonas aeruginosa*.

Table 1 A Comparative Study of Resonance Raman Between Transferrins and Dioxygenases

Protein	Principal Resonance Raman Peaks (cm^{-1})				Reference
Ovotransferrin/HCO$_3^-$	1605	1504	1270	1170	18
Serum transferrin/HCO$_3^-$	1613	1508	1288	1174	19
Ovotransferrin/C$_2$O$_4^{2-}$	1604	1505	1264	1175	20
PCA 3,4-oxygenase (*aeruginosa*)	1602	1503	1263	1171	16
PCA 3,4-oxygenase (*putida*)	1608	1504	1269	1175	15
Pyrocatechase	1605	1505	1293	1173	17

cipal resonance Raman peaks for three transferrins and three intradiol-cleaving dioxygenases.

3.2 PCA 4,5-Oxygenase

When *Pseudomonas testosteroni* is grown on *p*-hydroxybenzoic acid as sole carbon and energy sources, the enzyme protocatechuic acid 4,5-oxygenase is induced. This enzyme catalyzes the cleavage of the aromatic ring of PCA to give α-hydroxy-γ-carboxy-*cis,cis*-muconic semialdehyde as ring fission product with the consumption of one molecule of oxygen. Dagley and Patel were the first investigators to demonstrate that cleavage of protocatechuic acid occurred adjacent to the two hydroxyl groups to give a semialdehyde product (21). The metabolic pathway for the degradation of PCA was subsequently worked out by Dagley et al. (22). The cyclization of α-hydroxy-γ-carboxy-*cis,cis*-muconic semialdehyde with ammonium ions to give 2,4-lutidinic acid as product helped to establish that 4,5-fission of the aromatic ring had occurred.

PCA 3,4-oxygenase is a very stable enzyme; however PCA 4,5-oxygenase is extremely unstable (23, 24). In 1968 Dagley et al. purified PCA 4,5-oxygenase to homogeneity from extracts of *Ps. testosteroni* (25). The enzyme was found to have a molecular weight of approximately 140,000, and Fe(II) was found to be an absolute requirement to activate the enzyme. When diluted, PCA 4,5-oxygenase rapidly deactivates. Deactivation was shown to be partially prevented by the inclusion of 5 mM L-cysteine and 10% glycerol in buffer solutions (25).

Recently Lipscomb has shown that the native enzyme is dissociated in SDS to eight subunits with a molecular weight of 15,000 to 16,000. These subunits appear to be identical. Zabinski et al. (26) confirmed that enzyme preparations could be activated by Fe(II) and attempted to determine the oxidation state of the iron centers in PCA 4,5-oxygenase by using EPR and Mössbauer spectroscopy. Several difficulties were encountered in this study. Lipscomb has shown that the preparations of PCA 4,5-oxygenase used by Zabinski et al. are at least an order of magnitude lower in specific activity than enzyme preparations prepared by affinity chromatography. Secondly, substrates and inhibitors of PCA 4,5-oxygenase chelate both Fe(II) and Fe(III) readily, making it difficult to unambiguously assign iron in the active site. Spectral changes would be expected for substrates or competitive inhibitors reacting with active site iron, adventitiously bound iron, or low molecular weight iron complexes in solution. In future studies it will be important to correlate iron content with the specific activity of enzyme preparations. It should be pointed out that Fe(II) activation of PCA 4,5-oxygenase does not yield any information on the oxidation state of the iron at the active site. PCA 3,4-oxygenase is activated by Fe(II), but the active sites formed by reconstitution with Fe(II) are undoubtably high-spin Fe(III). Clearly a single electron transfer reaction from Fe(II) to PCA 3,4-oxygenase must accompany the formation of active high-spin Fe(III) centers.

3.3 PCA 2,3-Oxygenase

Crawford (27) was the first to demonstrate that cleavage of PCA can occur by an extradiol mechanism between carbons 2 and 3. The enzyme PCA 2,3-oxygenase has now been purified by affinity chromatography from extracts of *Bacillus macerans* grown with p-hydroxybenzoate as sole carbon source (28). The native enzyme has a molecular weight of 90,000 and can be dissociated on polyacrylamide into subunits of 30,000 and 15,000 daltons.

Sephadex G 100 chromatography of native enzyme leads to dissociation into subunits 45,000 in molecular weight. The appearance of subunits of three different sizes can be rationalized as follows. If we assume that the α subunits have a molecular weight of 15,000 and the β subunits a molecular weight of 30,000, then $\alpha_2 = 30,000$ and $\alpha\beta = 45,000$ daltons. PCA 2,3-oxygenase is activated by Fe(II) and deactivates upon dilution.

4 SUBSTRATE BINDING AND KINETICS

Native PCA 3,4-oxygenase has a λ_{max} at 450 nm with broad absorption over the range of 400 to 650 nm. When substrate is added under anaerobic conditions a shift to the red occurs (λ_{max} 480 nm) which is accompanied by a small increase in the extinction of the chromophore. Introduction of dioxygen to the enzyme–substrate complex (when 3,4-dihydroxyphenyl propionate is used as substrate) causes a shift to a new species with a λ_{max} at 520 nm. Based on this reaction sequence Fujisawa et al. (29) proposed the following reaction pathway:

$$E + S \rightleftharpoons ES + O_2 \rightarrow ESO_2 \rightarrow E + P$$

450 nm   480 nm   520 nm

Recently Nakata et al. (30) have revised this scheme as follows:

$$E + S \rightleftharpoons ES + O_2 \rightarrow ESO_2 \rightarrow EP \rightarrow E + P$$

450 nm   480 nm   520 nm

These workers now believe that the steady-state complex with a λ_{max} at 520 nm is undissociated EP complex.

Que et al. prepared the 450 nm species (E), the 480 nm species (ES), and the 520 nm species (ESO_2) as well as the enzyme–product complex (EP), and examined them by EPR and Mössbauer spectroscopy (31, 32). The native enzyme (E) shows prominent features at $g = 4.3$ and $g = 9.35$, typical of high-spin ferric iron in a "rhombic" environment. Blumberg and Peisach (33) conducted variable-temperature measurements on PCA 3,4-oxygenase and determined the zero-field splitting parameter D for this enzyme as 1.6 cm^{-1} (λ was determined as 0.28). This zero-field splitting parameter is similar to that of Fe(III)-rubredoxin and is much larger than the values found for other proteins exhibiting a $g = 4.3$ resonance. For example, high-spin Fe(III) complexes coordinatively saturated with oxygen have zero-field splitting parameters of $D \simeq 0.8$ cm^{-1}. On the basis of this initial EPR study Blumberg and Peisach suggested that PCA 3,4-oxygenase could have a ligand environment similar to that of rubredoxin: a tetrahedral environment with four cysteinyl–sulfur ligands. When substrate binds to PCA 3,4-oxygenase, the "rhombic" signal at $g = 4.3$ decreases, and "axial" signals around $g = 6.0$ appear. In all cases two sets of EPR signals appear upon substrate binding. These two sets of signals appear to be genuine ES complexes because they appear at less than saturating substrate concentrations. These ES complexes are characterized by a zero-field splitting parameter $D \simeq 0$ and $\lambda = 0.02$ and 0.12 for the two species. The introduction of oxygen into the ES mixture yields a ternary oxy complex, ESO_2, with signals at $g = 6.7$ and 5.3. These signals arise from high-spin ferric iron in an "axial" environment characterized by a negative zero-field splitting parameter $D = -2.0$ cm^{-1} ($\lambda = 0.03$). This complex has unique magnetic resonance properties among biological iron complexes.

SUBSTRATE BINDING AND KINETICS

The enzyme–product complex (EP) does not give EPR signals around $g = 6.0$. Thus the resonance at $g = 6.7$ cannot be associated with an EP complex; and since it is formed only upon the introduction of oxygen it must be some form of ternary enzyme–substrate–oxygen intermediate (31).

This magnetic resonance study complicates the interpretation of the stopped-flow experiments because it is clear that two ES complexes are formed. Furthermore the 520 nm complex must be an ESO_2 complex and not the EP complex because these EPR results rule out the possibility of this species being assigned as EP. On the basis of the results of Que et al. (31, 32), the following pathway seems more likely:

$$\begin{array}{c} ES + O_2 \\ \searrow \\ E + S \qquad\qquad ESO_2 \;\;\rightarrow\; EP \rightarrow E + P \\ \nearrow \\ ES' + O_2 \end{array}$$

The difference between ES and ES' may be the result of an equilibrium between a protonated and unprotonated enzyme–catecholate complex.

The Mössbauer spectra of PCA 3,4-oxygenase contain a great deal of valuable information on these unusual high-spin Fe(III) complexes (31). Mössbauer spectroscopy has established that the native enzyme, the ES complexes, and the ESO_2 complex all contain high-spin Fe(III) iron. Also, even though the native enzyme has an EPR spectrum similar to that of rubredoxin ($D = 1.6 \text{ cm}^{-1}$), the isomer shifts suggest more ionic, oxygen, and nitrogen ligands, rather than a tetrahedron of cysteinate residues. In fact the magnetic hyperfine coupling contribution is inconsistent with a rubredoxin type of environment, especially when oxidized rubredoxin and PCA 3,4-oxygenase are compared. Octahedral coordination by oxygen ligands is ruled out on the basis of the zero-field splitting parameter of $D = 1.6 \text{ cm}^{-1}$. Such complexes give a characteristic zero field splitting parameter D less than 1.0 cm^{-1}. A Mössbauer study of reduced PCA 3,4-oxygenase suggests octahedral coordination, but unusual tetrahedral coordination to the iron cannot be unambiguously ruled out (34). Reduced PCA 3,4-oxygenase has spectral features very similar to deoxyhemeythrin. Taken together, these data suggest that the presence of oxygen and nitrogen ligands to the iron probably yield an octahedral complex. However tetrahedrally coordinated ferric iron cannot be ruled out at this time.

Substrate analogs have been used to assess the relative importance of functional groups on substrate binding (32). In PCA 3,4-oxygenase the iron could bind to any one of the substituent groups, chelate to both hydroxyl groups, or not bind at all. Kinetic studies, as well as EPR and optical spectroscopic experiments, have been conducted for a series of substrate analogs where one of the functional groups is either modified or missing. The results of these kinetic experiments are presented in Table 2. It is clear from the inhibition data for analogs II, III, and IV that the carboxylate group is important for binding. All three compounds are substrates for the enzyme, though their turnover rates are much smaller than for protocatechuate (I); as such, they can be considered competitive inhibitors to the protocatechuate cleavage reaction. The fact that all three are pseudosubstrates suggests that the carboxylate group is not essential for activity. Furthermore, the optical absorption spectra of the enzyme with each of the three inhibitors are similar to the spectrum of the enzyme–protocatechuate complex; in all cases λ_{max} shifts from 450 nm in the native enzyme to 480 nm with a concomitant increase in the absorption at 600 nm. The above observations all suggest that while the carboxylate is important for binding, it does not bind to the iron.

The carboxylate binding site is most probably a positively charged moiety capable of electrostatic interaction with negatively charged groups, probably the ε-amino group of a lysyl residue. Compounds with carbonyl functional groups replacing the carboxylate (V, VI, and VII) can interact with this group and are indeed inhibitory to the enzymatic reaction. In particular, protocatechualdehyde (V) could react with the amino group to form a Schiff base which could then be reduced by sodium borohydride. This has been tested by Que et al., who used tritiated proto-

Table 2 Dissociation Constants for Protocatechuate 3,4-Dioxygenase

	Substrate Analog	$K_I{}^a$ (μM)	$K_D{}^b$ (μM)
II	3,4-Dihydroxyphenyl acetate	13	
III	3,4-Dihydroxyphenyl propionate	18	
IV	Catechol	700	
V	Protocatechualdehyde	14	
VI	3,4-Dihydroxyacetophenone	130	130
VII	Protocatechuic acid methyl ester	400	300
VIII	p-Hydroxybenzoate	240	180
IX	m-Hydroxybenzoate	4	
X	Vanillate	3	
XI	Isovanillate	20	
XII	3-Fluoro-4-hydroxybenzoate	1	
XIII	4-Fluoro-3-hydroxybenzoate	800	
XIV	2-Fluoro-4-hydroxybenzoate	1	

[a] Obtained from inhibition kinetic data.
[b] Obtained from optical titration experiments.

catechualdehyde (V) and succeeded in covalently labeling the α subunit by reducing the Schiff base to an enamine derivative (31).

Substrate analogs in Table 2 are grouped in pairs of meta and para isomers. These kinetic data clearly indicate that the p-hydroxy group is much more important to binding than the m-hydroxy group. The most striking difference is seen with the fluoro derivatives XII and XIII. The p-hydroxy derivatives XII binds three orders of magnitude better than XIII. The enzyme–inhibitor complexes of the two fluoro derivatives XII and XIII have been examined by EPR spectroscopy, and the results of this study are presented in Table 3. As with the substrate, two enzyme–inhibitor complexes EI and EI' are seen upon inhibitor binding. The zero-field splitting parameter D for the enzyme–XII complex is similar to that of the ESO_2 complex. This provides further evidence for the selective coordination of the p-hydroxy group to the high-spin ferric active site.

For the extradiol-cleaving enzymes PCA 4,5-oxygenase and PCA 2,3-oxygenase, detailed binding studies have not yet been done. However preliminary kinetic experiments with PCA 4,5-oxygenase implicate a lysine residue at the carboxylate binding site (35). There is no clear-cut evidence for the preferential binding of either the p-hydroxy or the m-hydroxy groups to the iron center of PCA 4,5-oxygenase.

5 DIOXYGENASE MECHANISMS

A number of mechanisms have been proposed for the dioxygenases. Most of the mechanisms postulate the interaction of a ferrous complex with

Table 3 EPR Parameters for Protocatechuate 3,4-Dioxygenase and its Complexes

Enzyme Complex	D (cm^{-1})	λ	μ
Native enzyme	1.6[a]	0.28	
Enzyme–substrate	>0	0.02 0.12	
Enzyme–substrate–O_2	−2.0	0.03	
Enzyme–XII	−1.72 −1.66	0.13 0.11	0.16
Enzyme–XIII	−1.08 −1.00	0.21 0.13	0.28

[a] See reference number 33.

dioxygen. For PCA 3,4-oxygenase it has been postulated that the substrate reduces the high-spin Fe(III) center to facilitate an interaction between oxygen and the iron center (2). In the insertion of oxygen a peroxy-type intermediate, XVI, is preferred over dioxetane XVII for thermodynamic reasons (36) (Scheme 5):

Scheme 5 Alternate mechanisms for oxygen insertion into catechol.

The discovery that PCA 3,4-oxygenase retains its high-spin Fe(III) character throughout the entire catalytic cycle has introduced some new thoughts on the mechanism of action of dioxygen insertion. To date, molecular oxygen has been found to bind only to high-spin ferrous iron complexes in the iron-containing enzymes. This led us to consider the possibility that the metal center may "activate" the organic substrate prior to oxygen insertion.

The reaction of catechols with oxygen is well documented. The initial products of catechol autoxidation as observed by EPR spectroscopy are the corresponding semiquinone (37) and presumably superoxide anion. Kinetic studies of this reaction show that the autoxidation rate of catechol increases with the extent of dissociation, the dianionic catecholate having a greater reactivity toward oxygen (38). Recently Que and Heistand have resonance Raman evidence that the dianionic catecholate binds to pyrocatechase (17). Grinstead (39) has observed that in fairly alkaline solution (pH 12 to 14), the reaction of oxygen with 3,5-di-t-butylcatechol yielded ring cleavage products as well as the corresponding quinone. Therefore the activation of the catechol to facilitate attack by molecular oxygen is not out of the question.

Let us consider a mechanism for PCA 3,4-oxygenase based on the kinetic and structural data outlined above. As the substrate enters the active site, the carboxylate end is anchored by an ϵ-amino group of a lysyl residue. The iron binds the 4-hydroxy group of the substrate and acts as a Lewis acid promoting electron migration toward C-4. This would lead to the keto resonance form of the catechol. The keto form of the catechol could react with dioxygen to give the semiquinone and superoxide. The superoxide is a better base than phenolate, and so it could coordinate to the Fe(III) center and the semiquinone radical, and the radical formed by the interaction of Fe(III) with superoxide could couple to give a peroxy intermediate (ESO$_2$). The resulting peroxy intermediate would be expected to rearrange to give an anhydride (EP) which would open by a pH-sensitive reaction to give *cis,cis*-muconic acid (P). This reaction pathway is presented in Scheme 6. The rearrangement of the peroxy intermediate to an anhydride is well known in organic chemistry. For example *trans*-decalin peresters rearrange to give anhydrides. This rearrangement is enhanced by electron-withdrawing groups and retarded by electron-donating groups (40).

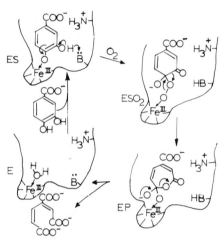

Scheme 6 Proposed reaction mechanism for PCA 3,4-oxygenase.

From this proposed mechanism it is clear that the oxygenation step requires electron donation to oxygen, and the rearrangement of the peroxy intermediate (ESO_2) to the anhydride (EP) requires electron withdrawal from the aromatic ring. Thus the nitro, the formyl, and the carbomethoxyl substituents are too electron withdrawing to allow reaction with oxygen, while the carboxyethyl and unsubstituted catechols react to form the peroxide intermediate. The latter substituents however are now too electron donating and consequently stabilize the peroxide intermediate and retard the rate of reaction. The carboxyl-substituted catechol is the only substrate with the appropriate electron balance to make efficient use of this enzyme. Table 4 shows the importance of this concept.

For the extradiol-cleaving enzymes PCA 4,5-oxygenase and PCA 2,3 oxygenase, we do not have sufficient information to propose mechanisms. These enzymes differ from PCA 3,4-oxygenase in a number of respects. Both PCA 4,5- and 2,3-oxygenases contain active centers which are extremely labile; enzyme activity is lost simply by diluting the protein with buffers. Although these enzymes can be activated by Fe(II) we do not know whether their active sites are Fe(II), Fe(III), or both. We do not

Table 4 Relative Rates of Reaction for Substituted Catechols

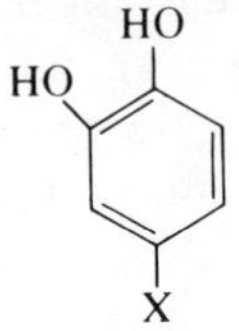

X—	Relative Rate	σ_p
—$CH_2CH_2COO^-$	0.02	−0.13
—H	0.4	0
—COO^-	100	+0.13
—COOMe	0	+0.39
—CHO	0	+0.45
—NO_2	0	+0.78

know whether the loss of activity upon dilution is caused by the loss of iron or the loss of an iron-containing cofactor. Preliminary studies indicate that these enzymes are inactivated by photolysis. PCA 4,5-oxygenase appears to have no preference for the coordination of either the *m*-hydroxy or *p*-hydroxy groups of competitive inhibitors. Could both phenolate groups coordinate at the active sites of the extradiol cleaving enzymes? We hope to answer some of these questions by constructing parallel studies with these three fascinating enzymes.

ACKNOWLEDGMENTS

The author wishes to thank his colleagues Drs. E. Münck, L. Que, J. D. Lipscomb, J. Howard, R. L. Crawford, W. H. Orme-Johnson, and D. Ballou for valuable discussions and to acknowledge the technical assistance of F. Engle, J. Bromley and R. L. Thrift. This work was supported by NSF Grants PCM 74–17318 and PCM 76–82502.

REFERENCES

1. O. Hayaishi and Z. Hashimoto, *J. Biochem. Tokyo*, **37**, 371 (1950).
2. T. Nakazawa, Y. Kojima, H. Fujisawa, M. Nozaki, and O. Hayaishi, *J. Biol. Chem.*, **240**, 3224 (1965).

REFERENCES

3. S. Dagley, *Am. Scientist*, **63**, 681 (1975).
4. S. Dagley, in *Essays in Biochemistry*, P. N. Campbell, Ed., Biochemical Society, 1976, pp. 81–137.
5. *Degradation of Synthetic Organic Molecules in the Biosphere*, National Academy of Sciences Report, Washington, D.C., 1972.
6. D. T. Gibson, J. R. Koch, and R. E. Kallio, *Biochemistry*, **7**, 2653 (1968).
7. R. L. Crawford, T. Frick, and P. Olson, *J. Bacteriol.*, in press (1979).
8. J. M. Wood, J. D. Lipscomb, L. Que, Jr., R. S. Stephens, W. H. Orme-Johnson, E. Münck, W. P. Ridley, L. Dizikes, A. Cheh, M. Francia, T. Frick, R. Zimmermann, and J. Howard, in *Biological Aspects of Inorganic Chemistry*, A. W. Addison, W. R. Cullen, D. Dolphin, and B. R. James, Eds., Wiley Interscience, New York, 1977, pp. 261–288.
9. R. Y. Stanier and J. L. Ingraham, *J. Biol. Chem.*, **210**, 799 (1954).
10. H. Fujisawa, M. Uyeda, T. Kojima, M. Nozaki, and O. Hayaishi, *J. Biol. Chem.*, **247**, 4414 (1972).
11. M. Nozaki, R. Yoshida, C. Nakai, M. Iwaki, Y. Saeki, and H. Kagamiyama, *Adv. Exp. Med. Biol.*, **74**, 127 (1976).
12. J. D. Lipscomb, J. Howard, T. Lorsbach, and J. M. Wood, *Fed. Proc.*, **35**, 1536 (1976).
13. N. Kohlmiller and J. Howard, *J. Biol. Chem.*, **00**, 0000 (1979).
14. Rao and Sundaralingham, personal communication.
15. D. Ballou and C. Bull, Isolation and Properties of Protocatechuate Dioxygenase from *Pseudomonas putida*. Proceedings of a Conference on Oxygen in Biochemistry. Pingree Park, Colorado, W. Caughey, Ed., 1979.
16. W. E. Keys, T. M. Loehr, and M. L. Taylor, *Biochem. Biophys. Res. Commun.*, **83**(3), 941 (1978).
17. L. Que, Jr., and R. Heistand III, *J. Am. Chem. Soc.*, **00**, 0000 (1979).
18. Y. Tomimatsu, S. Kint, and J. R. Scherer, *Biochemistry*, **15**, 4918 (1976).
19. B. P. Gaber, V. Miskowski, and T. G. Spiro, *J. Am. Chem. Soc.*, **96**, 6868 (1974).
20. P. R. Carey and N. M. Young, *Can J. Biochem.*, **52**, 273 (1974).
21. S. Dagley and M. D. Patel, *Biochem. J.*, **66**, 227 (1957).
22. S. Dagley, W. C. Evans, and D. W. Ribbons, *Nature*, **188**, 560 (1960).
23. M. L. Wheelis, N. J. Palleroni, and R. Y. Stanier, *Arch. Mikrobiol.*, **59**, 302 (1967).
24. K. Ono, M. Nozaki, and O. Hayaishi, *Biochim. Biophys. Acta*, **220**, 224 (1970).
25. S. Dagley, P. J. Geary, and J. M. Wood, *Biochem. J.*, **109**, 559 (1968).
26. R. M. Zabinski, E. Münck, P. M. Champion, and J. M. Wood, *Biochemistry* **11**, 3212 (1972).
27. R. L. Crawford, *J. Bacteriol.*, **121**, 531 (1975).
28. R. L. Crawford, P. Perkins-Olson, and J. Bromley, *Appl. Environ. Microbiol.*, in press.
29. H. Fujisawa, K. Hiromi, M. Uyeda, M. Nozaki, and O. Hayaishi, *J. Biol. Chem.*, **247**, 4422 (1972).
30. H. Nakata, T. Yamauchi, and H. Fujisawa, *Biochim. Biophys. Acta*, **527**, 171 (1978).
31. L. Que, J. D. Lipscomb, R. Zimmermann, E. Münck, N. R. Orme-Johnson, and W. H. Orme-Johnson, *Biochim. Biophys. Acta*, **452**, 320 (1976).
32. L. Que, J. D. Lipscomb, E. Münck, and J. M. Wood, *Biochim. Biophys. Acta*, **485**, 60 (1977).

33. W. E. Blumberg and J. Peisach, *Ann. N.Y. Acad. Sci.*, **222,** 539 (1973).
34. R. Zimmermann, B. H. Huynh, E. Münck, and J. D. Lipscomb, *J. Chem. Phys.*, **69**(12), 5462 (1978).
35. R. M. Zabinski, Ph.D. Thesis in Biochemistry, University of Illinois, Urbana, 1972.
36. G. A. Hamilton, in *Molecular Mechanisms of Oxygen Activation*, O. Hayaishi, Ed., Academic Press, New York, 1974, pp. 443–445.
37. M. Adams, M. S. Blois, Jr., and R. H. Sands, *J. Chem. Phys.*, **28,** 774 (1958).
38. M. A. Joslyn and G. E. K. Branch, *J. Am. Chem. Soc.*, **57,** 1779 (1935).
39. R. R. Grinstead, *Biochemistry*, **3,** 1308 (1964).
40. E. S. Gould, *Mechanism and Structure in Organic Chemistry*, Holt Rinehart and Winston, New York, 1959, p. 633.

CHAPTER 5
Cytochrome c Oxidase

BO G. MALMSTRÖM

Department of Biochemistry and Biophysics,
Chalmers Institute of Technology and
University of Göteborg, Göteborg, Sweden

CONTENTS

1 INTRODUCTION, 183

    1.1 The Reactions of Cytochrome c Oxidase and the Biological Distribution of the Enzyme, 183
    1.2 Brief Historical Survey and Scope of the Chapter, 184

2 THE PROSTHETIC GROUP OF THE OXIDASE AND OF ITS SUBSTRATE, 185

    2.1 Chemical Structure, 185
    2.2 Electron-Transfer Mechanism to the Heme Groups, 186

3 STRUCTURE OF CYTOCHROME c OXIDASE, 187

    3.1 Chemical Composition, 187
    3.2 The Peptide Subunits of the Oxidase, 188
    3.3 The Arrangement in the Membrane, 189

4 SPECTROSCOPIC AND MAGNETIC PROPERTIES, 191

    4.1 Visible and Infrared Spectra, 191
    4.2 EPR Spectra, 192
        4.2.1 The Heme Signals, 193
        4.2.2 The Copper Signal, 195
        4.2.3 Model Derived from EPR Properties, 195
    4.3 Magnetic Susceptibility, 196

5 OXIDATION–REDUCTION PROPERTIES, 197

    5.1 Redox Titrations Based on Optical Measurements, 197
    5.2 Interactions, 198

6 THE CATALYTIC REACTION, 199

    6.1 The Sequence of Electron Transfer, 199
    6.2 "Allosteric" Effects, 200
    6.3 Oxygen Intermediates, 201

7 CONCLUDING REMARKS, 202

ACKNOWLEDGMENTS, 203

REFERENCES, 203

1 INTRODUCTION

1.1 The Reactions of Cytochrome c Oxidase and the Biological Distribution of the Enzyme

Most of the dioxygen absorbed by aerobic organisms is consumed in the reaction catalyzed by cytochrome c oxidase:

$$4c^{2+} + O_2 + 4H^+ \rightarrow 4c^{3+} + 2H_2O \tag{1}$$

where c^{2+} and c^{3+} represent the reduced and the oxidized form, respectively, of cytochrome c. The reason for the quantitative dominance of this process of dioxygen uptake is the fact that it constitutes the terminal reaction of the so-called respiratory chain. This provides most of the free energy needed for the life processes of aerobic organisms by coupling electron transport to the synthesis of adenosine triphosphate, ATP. In view of its key role in oxidative phosphorylation, cytochrome oxidase has a wide biological distribution. It is found in all animals and plants, in aerobic yeasts, and in some bacteria. In all cases it is firmly associated with a membrane, either the inner membrane of mitochondria, the respiratory organelle of higher organisms, or, in bacteria, the cell membrane.

It is now established (1) that ATP synthesis in oxidative phosphorylation is driven by an electrochemical potential across the membrane, as first suggested by Mitchell in 1961 (2). This potential is generated by a coupling of the electron transport in the respiratory chain to the creation of a proton gradient across the membrane (1). It has long been known that one site of ATP synthesis is located between cytochrome c and dioxygen, and it has now been shown that cytochrome oxidase is directly involved in the generation of a membrane potential (3–5). The nature of the coupling between the reaction of eq. (1) and the formation of the proton gradient is at present (January 1979) a matter of controversy (5–7), but there is accumulating evidence (7, 8) that the oxidase functions as a proton pump. In addition, protons are of course consumed on the side of dioxygen reduction according to eq. (1).

The direct coupling between electron transport from cytochrome c to dioxygen and the generation of the electrochemical potential across the membrane means that eq. (1) does not correctly describe the reaction catalyzed by cytochrome oxidase, which instead should be written (7)

$$4c^{2+} + O_2 + 8H_M^+ \rightarrow 4c^{3+} + H_2O + 4H_C^+ \tag{2}$$

in which the subscripts M and C refer to the matrix and cytosol side, respectively, of the inner mitochondrial membrane. As this book is concerned with dioxygen activation, emphasis will however be placed on the four-equivalent reduction of dioxygen as described by eq. (1).

With solubilized oxidase in detergent solution, the absence of a membrane structure has the consequence that the reaction of eq. (1) is effectively dissociated from proton translocation, and most detailed structural and mechanistic information has been derived from studies on such solubilized preparations (9). Purified cytochrome oxidase in detergent solution shows a low specific activity however (10). It can be considerably activated by the addition of phospholipids (11, 12), suggesting that the functional conformation requires a membrane structure. Consequently it has been suggested (13) that the spectroscopic and redox properties as well as the kinetic behavior of the enzyme best be studied in artificial membranes. If the membrane is impermeable to protons, the steady-state reduction of O_2 will be slow, inasmuch as the reaction of eq. (2), unlike that of eq. (1), does not have an equilibrium position far to be right. One can however separate the reaction of eq. (1) from proton transport by the addition of suitable ionophores. The stimulation of mitochondrial respiration on addition of ionophores, so-called respiratory control, is a well-known phenomenon (1), and it has now been demonstrated also with artificial membranes containing cytochrome oxidase (4, 14).

While the purpose of this chapter is to give a description of the present state of knowledge concerning the catalysis by cytochrome oxidase of the electron transfer from cytochrome c to dioxygen, the preceding discussion indicates that it may be unwise even in this context to limit oneself to studies of the reaction of eq. (1) rather than that of eq. (2). This may be particularly true in relation to the "allosteric" phenomena discussed in Section 6.2, as proton translocation by the oxidase functioning as an ion pump must involve redox-linked conformational changes in the enzyme.

1.2 Brief Historical Survey and Scope of the Chapter

During the 1920s Keilin (15) carried out his pioneering studies on the cytochromes, but he did not at that time realize that the terminal oxidase belongs to this group of respiratory proteins. It was Warburg (16) who first suggested that the "Atmungsferment" contains iron. Particularly convincing was his determination (16) in 1929 of the photochemical action spectrum of the CO complex of reduced oxidase, showing that this was similar to the corresponding spectrum for hemoglobin. Warburg (17) introduced the concept of the respiratory chain in 1933 by suggesting that the "Atmungsferment" is the terminal component in a chain of cytochromes. This view was accepted by Keilin (18) in 1938, when he showed that cytochrome c can reduce the oxidase, previously called "indophenol

THE PROSTHETIC GROUP OF THE OXIDASE

oxidase" by him. As a consequence Keilin introduced the name commonly used for the enzyme at present, cytochrome c oxidase. He did not accept that it is a heme-containing protein however, instead emphasizing its copper content (15). It was not until 1939 that the views of Warburg and Keilin were reconciled, when Keilin (19) discovered that his cytochrome a, having an absorption line close to 600 nm, was heterogeneous and had a component, cytochrome a_3, which behaves as Warburg's "Atmungsferment." The term cytochrome a was retained for the component that does not react directly with dioxygen.

Keilin's concept of two separate cytochromes in the oxidase has been a subject of controversy (9, 20) until a few years ago. It is now firmly established however that the minimal functional unit of cytochrome oxidase contains two heme groups in different environments, cytochromes a and a_3, as well as two copper atoms with distinct properties, Cu_A and Cu_B, as shown in Figure 1. It is the purpose of this chapter to attempt an up-to-date description of the principal properties of these redox-active metal centers and their function in the activation and reduction of dioxygen with the formation of two molecules of water. It is not the intention however to provide an exhaustive bibliography of the cytochrome oxidase literature, and only a selected number of representative references are given. Readers requiring a more complete documentation are referred to several reviews (9, 20–22). Surveys emphasizing recent developments not covered by the articles quoted are being prepared for *BBA Reviews on Bioenergetics* (23, 24).

2 THE PROSTHETIC GROUP OF THE OXIDASE AND OF ITS SUBSTRATE

2.1 Chemical Structure

While cytochrome oxidase contains two distinct cytochromes (Fig. 1), only one type of heme, designated heme a, can be isolated from it (20).

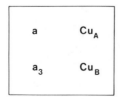

Figure 1. Prosthetic groups in the minimal functional unit of cytochrome c oxidase.

Thus the different properties of cytochromes a and a_3 must be derived from differences in the binding to the protein part of the enzyme. Cytochrome a has its iron atom in a low-spin state (Sections 4.2 and 4.3) and probably has two axial protein ligands, like other low-spin cytochromes, such as cytochrome c (25). Cytochrome a_3, on the other hand, appears to have a free coordination position, as it can interact with dioxygen or various inhibitors, and the iron atom is found to be in a high-spin state (Sections 4.2 and 4.3).

The structure of heme a as isolated by extraction is given in Figure 2 (26). The long side chain may be involved in binding the prosthetic group to the protein through hydrophobic interaction. It is also possible that its double bonds participate in electron transfer which then could occur over a considerable distance (cf. Section 2.2).

The heme in cytochrome c is covalently linked to the protein through two cysteine residues attached to positions 2 and 4 of the porphyrin ring. In addition the formyl group at position 8 of heme a (Fig. 2) is replaced by a methyl group. The protein provides two axial ligands, a N from histidine-18 and a S from methionine-80 (25). While most of the prosthetic group is embedded in a crevice of the protein, one edge is exposed to the solvent (25).

2.2 Electron-Transfer Mechanism to the Heme Groups

Dickerson et al. (27) have proposed a mechanism for the reduction of cytochrome c in which the electron hops from one aromatic residue to

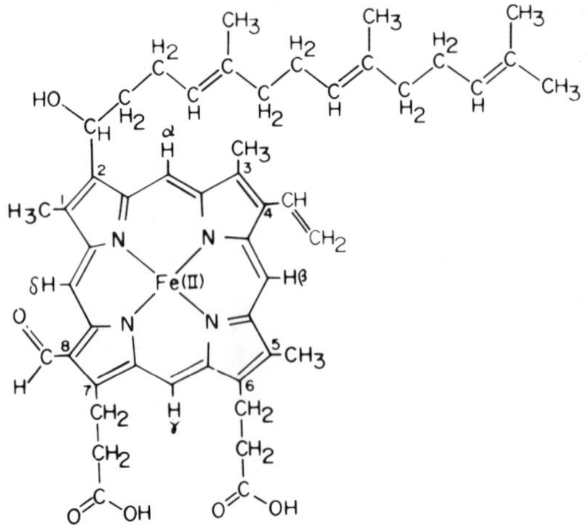

Figure 2. The chemical structure of heme a (26).

another, this hopping being facilitated by conformational changes. This appears unlikely however, as the free energy of activation for the addition of an electron to a tyrosine residue is large (28). In addition, tyrosine-74, which is the primary electron acceptor in this mechanism, is replaced by an aliphatic amino acid in some cytochromes (25).

Sutin (28) has given a discussion of other suggested reduction mechanisms. His own studies favor a direct electron transfer to the exposed heme edge. This does not however appear to be a likely mechanism for the electron transfer between cytochrome c and cytochrome a, the primary electron acceptor in the oxidase (Section 6.1), as the distance between the respective heme groups is thought to be about 2.5 nm (Section 3.3). Consequently a tunneling mechanism may have to be considered (29). Participation of the side chain may however allow electron transfer over considerable distances even by a classical mechanism (Section 2.1).

3 STRUCTURE OF CYTOCHROME c OXIDASE

3.1 Chemical Composition

The chemical composition of isolated cytochrome oxidase depends on the method used for solubilization and protein fractionation. For a discussion of the purification methods available, see the reviews quoted, particularly references 20 and 21. Recently some chromatographic procedures for the isolation of cytochrome oxidase have been described. One, involving affinity chromatography with Sepharose-bound cytochrome c (30), does not seem to be reproducible (31). Another method (32) utilizes hydrophobic interaction chromatography on octyl-Sepharose.

Most purified preparations of cytochrome oxidase contain 10 to 11 nmole heme a per mg protein, but some recent procedures (33, 34) result in a content of 13 to 14 nmole per mg. In terms of a functional unit with two hemes (Fig. 1), this corresponds to a protein molecular weight of at least 140,000. It may be noted that this value is about 20,000 higher than the sum of subunit molecular weights (Section 3.2) reported by some authors (35) but agrees closely with the subunit composition reported by Steffens and Buse (36). In addition it has been possible to crosslink the subunits to get a product having a molecular weight of 140,000 (37).

All oxidase preparations contain a considerable amount of lipids, mostly phospholipids (20). In most cases the phospholipid content is close to 20%, but some methods, for example hydrophobic interaction chromatography (32), remove the major part of the lipids (phospholipid content around 0.2%). Preparations with a low phospholipid content generally show an activity less than maximal, but they can be activated by the addition of phospholipids (10, 12, 13). The specificity of the phospholipid requirement appears low (12, 20), even if diphosphatidyl glycerol (cardiolipin) has been claimed to be essential (12). Most preparations seem however to contain enough bound diphosphatidyl glycerol for this requirement not to be limiting (12, 20, 21).

According to the generally accepted composition of the functional unit (Fig. 1), the copper-to-heme ratio should be 1.0; but it is generally closer to 1.1 (9). The extra copper is assumed to represent an impurity, a supposition which is supported by the fact that it gives a different EPR spectrum from the intrinsic Cu^{2+} (Section 4.2.2).

3.2 The Peptide Subunits of the Oxidase

Our knowledge of the biological synthesis and arrangement in the membrane of the subunits of cytochrome oxidase stems to a very large extent from work with the enzyme from yeast (38). Mechanistic investigations have on the other hand been carried out mostly with oxidase from beef heart. For this reason the discussion of structure will concentrate on the bovine oxidase. But as there appear to be many similarities with the yeast enzyme (35), some data for this are also included.

There has been some disagreement on the number of polypeptide chains of beef heart cytochrome oxidase, but several investigators (35, 39) have now reported that it, like the yeast enzyme (40), consists of seven subunits, as listed in Table 1. Steffens and Buse (36) claim however that subunit V consists of two polypeptides so that the actual number of subunits is eight. In addition subunit VIII appears heterogeneous. The subunits have been reported to be present in a 1:1 ratio (35, 39). Wikström et al. (41) have suggested a different stoichiometry on the basis of a presumed discrepancy between the sum of the subunit molecular weights

Table 1 Subunit Structure of Cytochrome c Oxidase (35)

Subunit	Molecular Weight
I	35,400
II	24,100
III	21,000
IV	16,800
V	12,400
VI	8,200
VII	4,400

and the molecular weight of the isolated enzyme. Recent data have however shown that with highly purified oxidase (high heme/protein ratio) there is a fairly good agreement between the two values (Section 3.1).

It is interesting to note that the three large subunits are coded for by mitochondrial DNA, while the synthesis of the remaining four is controlled by nuclear DNA (42). This is often used as one argument for a bacterial origin of mitochondria (43). In mutants lacking some mitochondrially synthesized subunits the cytoplasmically synthesized peptides are not assembled in the membrane (44). Also the heme, which is synthesized in the mitochondrion, is necessary for normal accumulation and assembly of the subunits (45).

Attempts have been made to identify the heme- and copper-binding subunits by isolation and analysis. Apparently the problem of redistribution is considerable, as almost all subunits have been found to bind the metals. Thus, peptides I, III, V, and VI have been said to be heme subunits (36, 39, 46, 47) and peptides II, V, and VII to contain copper (36, 39, 46). Steffens and Buse (36) suggest that the only safe identification at present must be based on sequence determinations and the finding of homologies with well-characterized heme and copper proteins. On such a basis they find subunit II to be a copper peptide, homologous to blue copper proteins, like stellacyanin, and VI (called VII by them) to be a heme subunit having a structure related to some cytochromes. Tanaka et al. (47) use the same reasoning to claim that subunit V binds heme.

Many experiments have also been designed to identify the subunit interacting with cytochrome c. Two groups find this to be subunit II (48–50), one subunit III (50), and one subunits VI and VII (52). One difficulty may stem from the length of the crosslinking reagents used in these investigations. Furthermore the identification of subunit III (51) was made with the yeast enzyme, and there is some doubt whether this corresponds to subunit III or II in the beef heart oxidase (48). The identification of bovine subunit II as a copper-binding peptide (36) makes it less likely however that this binds cytochrome c, as there is no dipolar interaction between the heme of cytochrome c and any paramagnetic center in the oxidase (K.-E. Falk, unpublished observations), meaning that the minimal distance between the metals is about 1.5 nm.

3.3 The Arrangement in the Membrane

The direct involvement of cytochrome oxidase in the creation of a membrane potential, whether by a redox loop mechanism (6) or by proton translocation through a proton pump (5), requires that the protein spans the lipid bilayer of the membrane. Electron microscopy studies (53, 54) with two-dimensional crystals have shown that the oxidase unit is asymmetrically placed in the membrane. Its total length is about 10 nm, approximately 4 nm of which protrudes above the lipid bilayer on the cytosol side and 1 nm below on the matrix side.

Several suggestions for the detailed arrangement of the subunits in the membrane have been published (48, 49, 51, 55). As many features of these models are very hypothetical, an additional, equally uncertain version is presented in Figure 3. This structure satisfies the more solid experiments, for example, the findings that subunits II, III, VI, and VII are exposed on cytosol side, subunit IV on the matrix side, while subunits I and V are inaccessible from either side (55). It does not however agree with all the conclusions from the more problematic crosslinking experiments. For example, Briggs and Capaldi (48) place peptide IV close enough to VI and VII to be crosslinked to these, a finding which it is hard to reconcile with the fact that these are small subunits located on opposite sides of the membrane. For reasons already stated, subunit II is not depicted as interacting directly with cytochrome c, but this is not excluded, as subunit II is large enough to put a copper ion 1.5 nm from the heme iron of bound cytochrome c. The possible location of one copper (Cu_A?) and one heme (cytochrome a?) has been indicated, while there is not good enough evidence to suggest binding sites for the other two metal centers (Section 3.2).

The orientation of the heme groups in relation to the membrane has been investigated by EPR and optical polarization spectroscopy (56–58). All results point to the heme normal lying in the membrane plane. Linear dichroism measurements (56) indicate that the oxidase molecule does not rotate along any axis in a time range up to 100 msec.

Fluorescence resonance energy transfer has been used to estimate the distances between a number of sites in the oxidase molecule (51). The heme of cytochrome c bound to subunit III was found to be 2.5 nm from the closest heme of the oxidase. It was also found to be 3.5 nm from a

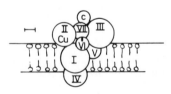

Figure 3. A possible arrangement of cytochrome c oxidase in the inner mithochondrial membrane. Bar represents 2 nm. Suggested locations of one heme (■) and one copper only are given. These, as well as many other features of the model, are highly uncertain and may need revision, as discussed in the text.

surface sulfhydryl group of subunit II, which in turn is 5.2 nm from the oxidase hemes (cf. Fig. 3).

4 SPECTROSCOPIC AND MAGNETIC PROPERTIES

4.1 Visible and Infrared Spectra

Like all heme proteins, cytochrome oxidase, both in its oxidized and in its reduced form, has several strong absorption bands in the spectral region of visible light. There is also weaker absorption in the infrared region. In Figure 4 the spectra for oxidized and reduced cytochrome oxidase in the region of 400 to 1000 nm are given. The Soret or γ-band between 400 and 470 nm is the most intense one with absorbance coefficients $>10^5$ M^{-1}/cm. The α-band around 600 nm is about an order of magnitude weaker, and the infrared absorption even lower. In the oxidized enzyme there is a distinct shoulder around 655 nm.

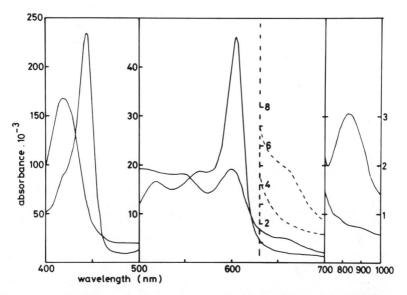

Figure 4. The absorption spectrum of cytochrome c oxidase. The spectrum has been recorded by B. Bejke, the author on Göteborg preparation 30 B, and gives slightly different absorbance coefficients at some wavelengths than those generally assumed.

For a number of reasons it is difficult to assign specific parts of the spectrum to given components of the oxidase. The bands are broader than those of proteins with a single heme, such as cytochrome c, indicating that both hemes contribute but have different absorption maxima, in agreement with the concept that they are in distinct chemical environments. In addition, while the heme bands are very strong, the contributions of the copper ions may not be negligible. For example, there is evidence (Sections 3.2 and 4.2.2) that Cu_A is a Type 1 Cu^{2+} (59), in which case it should have an absorbance coefficient of about 5000 M^{-1}/cm close to 600 nm. The infrared band at 830 nm is generally ascribed to copper (60), but there is disagreement about the relative absorption of Cu_A and Cu_B (61, 62), and in addition there is probably heme contributions in this region also (63).

Attempts at determining the relative contributions to the absorption spectrum of the different metals are usually based on differential reduction experiments (20). For example, in the presence of CO it is possible to reduce and oxidize cytochrome a while cytochrome a_3 stays reduced. Similarly at a chosen oxidation-reduction potential one can achieve a reduction of what appears to be one heme only (64). A major difficulty with all such experiment is however that they assume that the four metals act as independent centers in taking up electrons or binding ligands. There is now ample evidence that this is an unjustifiable simplification (24, 65; see also Sections 5.2 and 6.2). Consequently redox and kinetic measurements based on changes in optical absorption can seldom be related to specific sites, and it is therefore unfortunate that such experiments still dominate the field.

For a time it was believed that the 655 nm absorption is specific for oxidized cytochrome a_3 (66), but later experiments (67) made this idea untenable. It is however still likely that it represents a specific form of oxidized a_3 (67), perhaps the state in which it is antiferromagnetically coupled to Cu_B (Sections 4.2.3 and 4.3).

Resonance Raman measurements may be more discriminating, and changes ascribed specifically to cytochrome a or a_3 have been described (68). Magnetic circular dichroism (MCD) spectra are also to some extent more specific, as is exemplified briefly in Sections 4.2 and 4.3.

4.2 EPR Spectra

The EPR spectrum of oxidized cytochrome oxidase, and of the partially reduced enzyme at two different pH values, is shown in Figure 5. The spectrum of the oxidized enzyme is dominated by contributions from a low-spin Fe^{3+} with lines at $g=$ 3.0, 2.2, and 1.5 and a narrow signal at

SPECTROSCOPIC AND MAGNETIC PROPERTIES

$g = 2$, ascribed either to Cu^{2+} or a radical (Section 4.2.2), which largely hides the $g = 2$ line of the heme signal. On partial reduction a high-spin heme signal at $g = 6$ appears. At high pH this is diminished, but instead a new low-spin signal with lines at $g = 2.6$, 2.16, and 1.86 is observed.

4.2.1 The Heme Signals. The heme signal in the oxidized enzyme (Fig. 5) can be satisfactorily simulated on the assumption that it is due to a magnetically isolated low-spin Fe^{3+}, as shown in Figure 6 (69). This means that it is more than 1 nm away from any other paramagnetic center. Quantitation shows that it corresponds to one heme per functional unit (69), demonstrating that the two hemes are in different environments already in the oxidized, resting enzyme. The low-spin signal can be attributed to cytochrome a (9).

The $g = 6$ signal appearing on partial reduction is generally assigned to cytochrome a_3 (9, 69), though there have been suggestions (70, 71) that it represents cytochrome a. The conventional assignments are supported by the separate functions of the two cytochromes. Thus cytochrome a is an electron transfer agent and would be expected to have a heme coordination similar to cytochrome c. Cytochrome a_3, on the other hand, should have a free coordination site to be able to bind a molecule of dioxygen (cf. Section 2.1) and would be expected to have a high-spin heme as in deoxyhemoglobin and deoxymyoglobin.

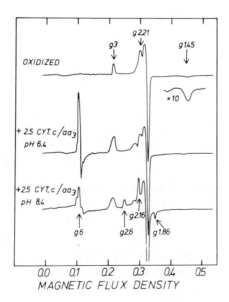

Figure 5. EPR spectra of oxidized (top) and partially reduced cytochrome c oxidase at pH 6.4 (middle) or 8.4 (bottom).

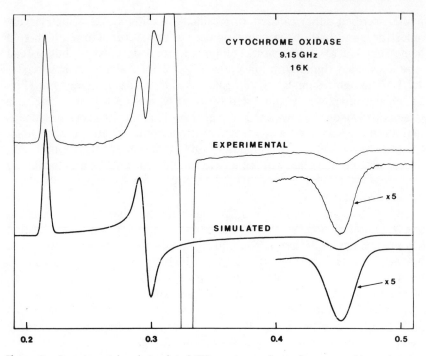

Figure 6. Experimental and simulated EPR spectrum of cytochrome c oxidase, showing the low-spin signal in detail (69).

The new low-spin signal appearing at high pH probably also represents cytochrome a_3 (72, 73). In Figure 7 it is shown that this signal represents the same amount of heme as corresponds to the decrease in the $g = 6$ intensity. A reasonable interpretation of these findings is that a proton dissociates from a water molecule bound to the Fe^{3+}, resulting in a metal-linked hydroxyl ion. The formation of metmyoglobin hydroxide at high pH is accompanied by a high- to low-spin conversion (74). This compound has g values (2.57, 2.14, and 1.84) that are strikingly similar to those of the high-pH signal of cytochrome oxidase.

The intensity of the $g = 6$ signal generally corresponds to <0.5 heme per functional unit (69, 73, 75). Recently Beinert and Shaw (76) have however prepared a form of oxidized cytochrome oxidase in which the major portions of both hemes are EPR detectable, one as a low- and one has a high-spin Fe^{3+}. This definitely rules out that the $g = 3$ and $g = 6$ signals represent different forms of the same heme component (at least in this particular compound).

MCD measurements (77, 78) have shown that the resting oxidase contains one high-spin and one low-spin heme, in agreement with the interpretation given here of the EPR results.

The parameters of the high-spin signal depend on experimental conditions (72, 73). This question will be further considered in Section 6.2.

4.2.2 The Copper Signal. The strong signal at $g = 2$ (Fig. 5) is generally ascribed to Cu_A (Fig. 1), but as there is no resolved hyperfine structure other interpretations are not excluded (69). Narrow hyperfine splittings and low g values are generally associated with so-called Type 1 Cu^{2+} (59). The parameters for the cytochrome oxidase signal are however extreme compared to other "blue" proteins. The signal is unusual also in terms of its relaxation properties. Thus it cannot be saturated at commonly available microwave powers, and it becomes too broad to be observed at temperatures above 150°K. The anomalous properties have led to the suggestion that it arises not from Cu^{2+} but from a radical structure, $Cu^+ \cdot S$ (79). Such a center is still related to Type 1 Cu^{2+}, which is coordinated to $-S^-$ with strong charge transfer from this ligand to the metal ion.

Most preparations of cytochrome oxidase have a minor component in their EPR spectra with parameters more normal for Cu^{2+} (69). This can be removed by chelating agents and is generally regarded as representing extraneous copper. A puzzling finding (69, 75) however is that the remaining "intrinsic" signal has an integrated intensity corresponding to 0.7 to 0.9 Cu^{2+} per functional unit only. In addition it has been found that even the so-called intrinsic or functional Cu^{2+} is heterogeneous (69).

4.2.3 Model Derived from EPR Properties. The EPR signals of oxidized cytochrome oxidase (Fig. 5) correspond at most to one Cu^{2+} and one

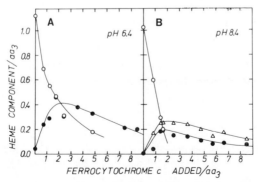

Figure 7. Reductive titrations of cytochrome c oxidase at ph 6.4 and 8.4 followed by EPR. The amounts of the different heme signals are $g = 1.45$ (0), $g = 6$ (●), and $g = 1.86$ (△) (73).

Fe^{3+}, that is, two of the four metals of the functional unit (Fig. 1). Still the EPR-invisible ions are also expected to be in their oxidized states, as the enzyme can accept four electrons anaerobically (75). The absence of EPR signals can be explained if cytochrome a_3 and Cu_B form an antiferromagnetically coupled pair, as first suggested by van Gelder and Beinert (80). Such magnetically coupled metals are found also in other oxidases capable of reducing dioxygen all the way to water, for example, laccase (59). Unlike laccase however, the two-electron reduction of this unit in cytochrome oxidase is not a cooperative process. Thus the $g = 6$ signal arises from those molecules which have cytochrome a_3 oxidized and Cu_B reduced. The nonintegral value of the maximum amount of oxidized a_3 formed would be explained by the metals not differing greatly in reduction potentials (cf. Section 5).

The EPR-detectable centers are cytochrome a and Cu_A, as already discussed (Sections 4.2.1 and 4.2.2). Cu_A is often assumed (81) to be associated with cytochrome a, probably because both components are reduced rapidly by cytochrome c (Section 6.1). This does not necessitate physical proximity however, and in fact the distance between these centers must be greater than 1 nm (Section 4.2.1). In analogy with laccase, which appears to have three metals in its dioxygen-reducing unit (82), it may instead be associated with the EPR-nondetectable ions.

4.3 Magnetic Susceptibility

The model based on EPR properties is supported by MCD (77, 78) as well as by magnetic susceptibility measurements. Until recently there were no satisfactory susceptibility data available for cytochrome oxidase, but now results at room temperature (83) as well as at low (7 to 200°K) temperatures (84) have been described. The room-temperature data are presented in Table 2. They show that the observed susceptibility of the ox-

Table 2 Magnetic Susceptibility of Cytochrome c Oxidase at 293°K (83)

Sample	Spin State of				$\Delta\chi_{mol} \times 10^6$
	a	Cu_A	a_3	Cu_B	
Oxidized	Observed				160
	$\frac{1}{2}$	$\frac{1}{2}$		2	160
Reduced	Observed				110
	0	0	2	0	130

idized enzyme can be accounted for by the presence of two $S = \frac{1}{2}$ centers, cytochrome a and Cu_A, and one $S = 2$ center, the antiferromagnetically coupled cytochrome a_3–Cu_B pair. The fact that this pair is coupled even at 293°K means that the exchange integral $J > 200$ cm^{-1}.

The reduced enzyme is still paramagnetic with $S = 2$. The simplest interpretation of this observation is that cytochrome a_3 remains high spin on reduction and that cytochrome a stays low spin, having $S = 0$ in the reduced state.

5 OXIDATION–REDUCTION PROPERTIES

5.1 Redox Titrations Based on Optical Measurements

On the phenomenological level there is rather general agreement on the redox properties of cytochrome oxidase as monitored by optical measurements. An illustrative example of Nernst plots, taken from the work of the Amsterdam group (85), is given in Figure 8. Measurements at 830 nm yields a straight line, with n in the Nernst equation equal to 1. The most natural interpretation of this finding is that at this wavelength one monitors the oxidation–reduction of a one-electron center. This would be in agreement with the common assumption that the 830 nm absorbance is mainly due to Cu_A^{2+} (Section 4.1).

The results based on changes in heme absorption at 605 or 405 nm are more complicated, yielding a sigmoid Nernst plot (Fig. 8). One way to

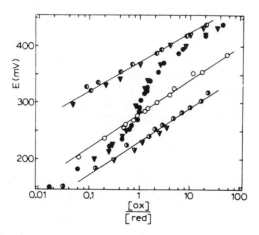

Figure 8. Potentiometric titration of cytochrome c oxidase followed at 445 nm (●), 605 nm (▼), and 830 nm (○) (85).

interpret this is illustrated by the resolved lines in Figure 8. They are based on the assumption that there are two separate one-electron centers with different reduction potentials but both contributing equally to the absorbance change at 405 as well as at 605 nm, one having a potential of 370 mV and the other a potential of 230 mV.

Unfortunately the conventional interpretation just summarized is not a unique one. It has been pointed out (9, 86) that the Nernst equation is formally identical with the Hill equation used to test cooperativity in a protein having several ligand-binding sites. It was in fact demonstrated (9) that the results in Figure 8 can equally well be described in terms of two hemes with the same potential but displaying a negative cooperativity of -100 mV. It was also shown that a straight line, as that obtained at 830 nm, can involve more than one redox center if there is a particular set of interactions. Originally these analyses were presented not to question the simple, noninteracting model but to point out that available data were insufficient to establish its validity. Subsequent investigations have however clearly shown that there are many manifestations of site–site interactions in the redox reactions of cytochrome oxidase.

5.2 Interactions

At the concentrations commonly used in stopped-flow kinetic experiments, the reaction of reduced cytochrome c with oxidized cytochrome a is much more rapid than subsequent reductive steps, resulting in a burst of cytochrome c oxidation (63, 87). Thus a quasi-equilibrium is established in the burst phase, allowing an estimate of the reduction potential for cytochrome a. The potential determined in this way is 285 mV (87) and thus does not correspond to any of the values found in the optical redox titrations (Section 5.1). If on the other hand a similar experiment is carried out with reduced cytochrome oxidase and oxidized cytochrome c, it is found that one heme site has a potential of 225 mV, a value close to that found for the low-potential heme in the experiment of Figure 8. These results can be summarized in the scheme of eq. (3):

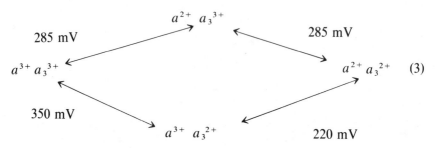

Thus both hemes are assumed to have potentials which depend on the redox state of the other heme. Similar schemes have also been derived from redox equilibrium measurements based on MCD and EPR changes (71) or on optical and EPR results (22).

The experiments discussed so far in this section refer only to reduction potentials of the two heme sites, and consequently the treatment has been limited to interactions between these. Recently a more complete analysis of interactions in a four-electron system has been applied to data from optical and EPR redox titrations available in the literature (65). It was found that in no case could a noninteracting model adequately describe the results. On the other hand, if interactions were introduced, then available data are not sufficient to discriminate between several possible models.

Apart from making it impossible at present to define microscopic equilibrium constants for the redox reactions of the four sites in cytochrome oxidase, the occurrence of interactions between these sites may mean that reduction potentials determined at equilibrium, as in the experiment of Figure 8, are completely uninteresting in terms of catalytically active species. Still, any mechanistic model based on kinetic measurements must also be consistent with the information derived from a complete analysis of the redox equilibrium studies.

6 THE CATALYTIC REACTION

Cytochrome oxidase, as well as the other enzymes reducing dioxygen to water (59), is asymmetric in the sense that electrons enter the enzyme from the reducing substrate at one site while dioxygen reacts at a different site (9, 88). Here present knowledge about the reactions at these two sites as well as about the electron transfer between them is summarized.

6.1 The Sequence of Electron Transfer

Much evidence suggests that the sequence of electron transfer from reduced cytochrome c to dioxygen can be described as in eq. (4):

$$c \to a \to Cu_A \to (Cu_B, a_3) \to O_2 \qquad (4)$$

Thus one mole oxidized cytochrome c per mole functional unit is formed in the burst (see Section 5.2) in a quasi-first-order reaction concomitantly with changes in the heme absorption of the oxidase (63, 87). (Reports that more than one oxidase site reacts in the burst are based on indirect experiments requiring knowledge which is in fact not available about the optical properties of the oxidase sites.) Furthermore additional cytochrome c is oxidized without corresponding changes in heme absorption (63, 89). It has also been found that the EPR signal assigned to Cu_A is diminished rapidly in reductive experiments (67). In addition carbon monoxide sensitivity, which depends on cytochrome a_3 being reduced, sets in late in the reductive sequence (89). (There is some evidence (62), unfortunately complicated by autoreduction, that both Cu_B and cytochrome a_3 need to be reduced for CO to bind.)

Finally the time course of reoxidation of reduced oxidase by dioxygen as measured at 605 nm is that shown in Figure 9 (90). It is not likely that the interpretation of this result is complicated by the interactions described in Section 5.2, as kinetically the different sites undoubtedly react in a specific sequence. Thus the initial rapid decrease in absorbance should reflect oxidation of cytochrome a_3, perhaps together with Cu_B. The subsequent inflection in the curve must be due to an absorbance increase during this phase of the reaction. This would be consistent with reoxidation of Cu_A, as a Type 1 Cu^{2+} absorbs strongly at 605 nm (Section 4.2.2), but it cannot be excluded that an intermediate in the reoxidation also absorbs at this wavelength (Section 6.3).

6.2 "Allosteric" Effects

The fact that the redox-active sites of cytochrome oxidase display a variety of interactions (Section 5.2) should also manifest itself by a type of

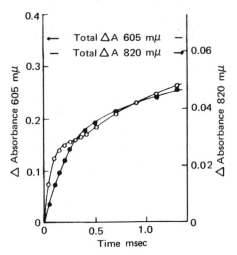

Figure 9. The reoxidation of reduced cytochrome c oxidase by dioxygen followed at 605 nm (○) and 820 nm (●) (90).

allosteric effects in kinetic experiments. Indeed already some very early observations suggest such phenomena. Thus in the oxidized resting enzyme, oxidized cytochrome a_3 does not react with cyanide, sulfide, or azide, although it is this heme component that binds these ligands (20). Binding occurs readily in inhibition experiments however, in which the presence of the reducing substrate leads to the reduction of other sites (see, for example, reference 91). More recently a large number of observations show that the resting oxidase represents a rather inactive conformation which is converted to a more active form when the enzyme is reduced and reoxidized.

It has been mentioned (Section 4.2.1) that the $g = 6$ signal varies with experimental conditions. Thus an axial signal is formed rapidly in reductive experiments but is slowly replaced by a more rhombic signal (67, 92). The rapidly formed signal responds slowly to dioxygen while the rhombic signal disappears rapidly on addition of reduced cytochrome c as well as of dioxygen. The rapidly responding signal can also be formed by reoxidation of reduced oxidase (72, 92). It is thus likely that the observations described are related to the so-called pulsed oxidase (93), an active form which is prepared by exposing reduced oxidase to a pulse of dioxygen. An EPR signal specific for the active enzyme form has been described (94). It was also suggested (94) that this form may be related to the classical "oxygenated" oxidase (20). Some results indicate however that the enzyme need not be fully reduced before oxidation to yield the active species (95).

A simple model to account for the results discussed would involve an equilibrium between two enzyme conformations:

$$R \text{ (inactive)} \rightleftharpoons T \text{ (active)} \qquad (5)$$

The two forms in eq. (5) could also account for some of the apparent heterogeneities of the enzyme. For example, it has been mentioned (Section 4.2.2) that the copper signal consists of two components. Only one of these responds rapidly to cytochrome c, and this could represent Cu_A in T while the other component is due to the same center in R. An equilibrium like that in eq. (5) could also be related to the changes in catalytic and EPR properties induced by the addition of phospholipid or decrease in pH (73).

6.3 Oxygen Intermediates

The great rapidity of the reaction of reduced cytochrome oxidase with dioxygen (90) has made the search for intermediates difficult. In the last few years considerable progress has however been made by B. Chance and associates with the use of low temperatures (down to 150°K) to slow down the reaction (96–98). Most work has been carried out starting with the reduced oxidase–carbon monoxide complex, which is stable even in the presence of dioxygen at the low temperature. The reaction is initiated by the carbon monoxide being dissociated with the use of flash photolysis. Some studies (97) have also been made with the so-called mixed valence oxidase, in which carbon monoxide is thought to be bound to reduced cytochrome a_3 and Cu_B^{1+}, while cytochrome a and Cu_A are oxidized (62). At least three intermediates, called compounds A, B, and C, have been found. These observations have been confirmed and extended by Clore and Chance (99).

The intermediates have been well characterized spectrophotometrically and kinetically (97, 99), but only incomplete EPR data are available (97). This has the consequence that it is difficult to assign unique electronic configurations, and some of the compounds may not even be true oxygen intermediates, in the sense that they represent dioxygen in different stages of reduction, but rather reflect different forms of the oxidase. With this reservation in mind it is still interesting to consider likely configurations.

Compound A is formed with a rate that is directly proportional to the concentration of O_2 and to an extent that can be described by a mass action relation for a 1:1 complex with a dissociation constant of 320 μM at 179°K (96). Thus it is likely that compound A is the true oxygenated oxidase (cytochrome $a_3^{2+} \cdot O_2$), but it cannot be excluded that electron transfer has occurred to yield $a_3^{3+} \cdot O_2^-$. Such a transfer is thermodynamically unfavorable however (9, 21, 100). Thus it seems more like that $a_3^{2+} Cu_B^{1+}$, which form an interacting unit, participate in a concerted two-electron donation to yield peroxide as the first reduced intermediate. Indeed compound B has been suggested to have the configuration $a^{2+} Cu_A^{1+} Cu_B^{2+} a_3^{3+} \cdot O_2^{2-}$ (97); an alternative interpretation (99) also has dioxygen at the peroxide stage. The configuration of the third intermediate is most uncertain (96, 99). A disturbing fact is also the finding that the intermediates formed from the mixed-valence state do not correspond to any of those produced from the fully reduced enzyme (97). Thus it is uncertain which intermediates are relevant to the true turnover condition with both reducing substrate and dioxygen present.

7 CONCLUDING REMARKS

Much of the large amount of work on cytochrome oxidase published during the 1970s has not brought any fundamental new insights but rather put on a firmer basis a model which was already available (9, 20). In this model electrons are transferred from cytochrome c to a low-spin heme component, cytochrome a, finally reaching the cytochrome a_3–Cu_B unit via Cu_A. Some things have been added to the model however. For example, it is now clear that redox-linked conformational changes occur, and these affect the rates of specific electron transfer steps (Section 6.2).

Antiferromagnetically coupled metal pairs have now been established as components in all oxidases reducing dioxygen to two molecules of water (101, 102). It is likely that their role is to facilitate a two-electron reduction directly to peroxide (103), thereby avoiding the thermodynamic barrier to the formation of O_2^- (Section 6.3). With laccase the presumed peroxide intermediate has been shown to decompose in one-electron steps, as a paramagnetic intermediate is formed (104). The use of oxygen isotopes has demonstrated that this is a radical, probably O^- (or OH) (82, 105). This is likely to be formed also in cytochrome oxidase, but no positive results with oxygen isotopes have been reported so far (72). Thus our understanding of the mechanism of dioxygen reduction in cytochrome oxidase is still very incomplete. In addition, there is hardly any information on the *molecular level* concerning the coupling of the electron transfer processes discussed in this chapter to proton translocation across the inner mitochondrial membrane, as indicated by (2) (Section 1.1). Cytochrome c oxidase may therefore be expected to be a fruitful object for mechanistic investigations for many years to come.

ACKNOWLEDGMENTS

This chapter was prepared while the author was a Visiting Professor in the Bioorganic Laboratory of the University of Utrecht, and he is indebted to Dr. J. F. G. Vliegenthart and his colleagues for facilities and stimulating discussions. The investigations in the author's home laboratory have been supported by grants from Statens Naturvetenskapliga Forskningsrad.

REFERENCES

1. P. D. Boyer, B. Chance, L. Ernster, P. Mitchell, E. Racker, and E. C. Slater, *Ann. Rev. Biochem.*, **46**, 955 (1977).
2. P. Mitchell, *Nature*, **191**, 144 (1961).
3. P. C. Hinkle, J. J. Kim, and E. Racker, *J. Biol. Chem.*, **247**, 1338 (1972).
4. D. R. Hunter and R. A. Capaldi, *Biochem. Biophys. Res. Commun.*, **56**, 623 (1974).
5. K. Krab and M. Wikström, *Biochim. Biophys. Acta*, **504**, 200 (1978).
6. J. Moyle and P. Mitchell, *FEBS Lett.*, **88**, 268 (1978).
7. M. Wikström and K. Krab, *FEBS Lett.*, **91**, 8 (1978).

8. E. Sigel and E. Carafoli, *Eur. J. Biochem.*, **89**, 119 (1978).
9. B. G. Malmström, *Q. Rev. Biophys.*, **6**, 389 (1973).
10. W. H. Vanneste, M. Ysebaert-Vanneste, and H. S. Mason, *J. Biol. Chem.*, **249**, 7390 (1974).
11. D. G. McConnell, A. Tzagoloff, D. H. MacLennan, and D. E. Green, *J. Biol. Chem.*, **241**, 2373 (1966).
12. C.-A. Yu, L. Yu, and T. E. King, *J. Biol. Chem.*, **250**, 1383 (1975).
13. B. Karlsson, B. Lanne, B. G. Malmström, G. Berg, and R. Ekholm, *FEBS Lett.*, **84**, 291 (1977).
14. F. B. Hansen, M. Miller, and P. Nicholls, *Biochim. Biophys. Acta*, **502**, 385 (1978).
15. D. Keilin, *The History of Cell Respiration and Cytochrome*, Cambridge University Press, Cambridge, 1966.
16. O. Warburg, *Schwermetalle als Wirkungsgruppen von Fermenten*, Editio Cantor, Freiburg, 1949.
17. O. Warburg, E. Negelein, and E. Haas, *Biochem. Z.*, **266**, 1 (1933).
18. D. Keilin and E. F. Hartree, *Proc. Roy. Soc.*, **B125**, 171 (1938).
19. D. Keilin and E. F. Hartree, *Proc. Roy. Soc.*, **B127**, 167 (1939).
20. M. R. Lemberg, *Physiol. Rev.*, **49**, 48 (1969).
21. W. S. Caughey, W. J. Wallace, J. A. Volpe, and S. Yoshikawa, in *The Enzymes*, Vol. 13, Part C, P. D. Boyer, Ed., Academic Press, New York, 1976, p. 299.
22. P. Nicholls and B. Chance, in *Molecular Mechanisms of Oxygen Activation*, O. Hayaishi, Ed., Academic Press, New York, 1974, p. 479.
23. M. Wikström and K. Krab, *Biochim. Biophys. Acta*, **549**, 177 (1979).
24. B. G. Malmström, *Biochim. Biophys. Acta*, **549**, 281 (1979).
25. R. E. Dickerson and R. Timkovich, in *The Enzymes*, Vol. 11, P. D. Boyer, Ed., Academic Press, New York, 1975, p. 397.
26. W. S. Caughey, G. A. Smythe, D. H. O'Keeffe, J. E. Maskasky, and M. L. Smith, *J. Biol. Chem.*, **250**, 7602 (1975).
27. R. E. Dickerson, T. Takano, O. B. Kallai, and L. Samson, in *Structure and Function of Oxidation–Reduction Enzymes*, Å. Åkesson and A. Ehrenberg, Eds., Pergamon, Oxford, 1972, p. 69.
28. N. Sutin, in *Bioinorganic Chemistry*—II, K. N. Raymond, Ed., American Chemical Society, Washington, D.C., 1977, p. 156.
29. R. H. Austin, K. W. Beeson, L. Eisenstein, H. Frauenfelder, and I. C. Gunsalus, *Biochemistry*, **14**, 5355 (1975).
30. T. Ozawa, M. Okumura, and K. Yagi, *Biochem. Biophys. Res. Commun.*, **65**, 1102 (1975).
31. S. Rosén, *Chromatographic Purification and Functional Properties of Cytochrome c Oxidase*, Ph.D. Thesis, University of Göteborg, 1978.
32. S. Rosén, *Biochim. Biophys. Acta*, **523**, 314 (1978).
33. C. R. Hartzell and H. Beinert, *Biochim. Biophys. Acta*, **368**, 318 (1974).
34. H. Komai and R. A. Capaldi, *FEBS Lett.*, **30**, 273 (1973).
35. N. W. Downer, N. C. Robinson, and R. A. Capaldi, *Biochemistry*, **15**, 2930 (1976).
36. G. F. Steffens and G. Buse, in *Cytochrome Oxidase*, T. E. King, Y. Orii, B. Chance, and K. Okunuki, Eds., Elsevier/North Holland, Amsterdam, 1979, p. 79.

REFERENCES

37. M. M. Briggs and R. A. Capaldi, *Biochemistry*, **16**, 73 (1977).
38. G. Schatz and T. L. Mason, *Ann. Rev. Biochem.*, **43**, 51 (1974).
39. C.-A. Yu and L. Yu, *Biochim. Biophys. Acta*, **495**, 248 (1977).
40. R. O. Poyton and G. Schatz, *J. Biol. Chem.*, **250**, 752 (1975).
41. M. Wikström, H. Saari, T. Penttilä, and M. Saraste, in *Membrane Proteins*, P. Nicholls, J. V. Møller, P. L. Jørgensen, and A. J. Moody, Eds., Pergamon Press, Oxford, 1978, p. 85.
42. T. L. Mason and G. Schatz, *J. Biol. Chem.*, **248**, 1355 (1973).
43. L. Sagan, *J. Theoret. Biol.*, **14**, 225 (1967).
44. F. Cabral and G. Schatz, *J. Biol. Chem.*, **253**, 4396 (1978).
45. J. Saltzgaber-Müller and G. Schatz, *J. Biol. Chem.*, **253**, 305 (1978).
46. S. Gutteridge, D. B. Winter, W. J. Bruyninckx, and H. S. Mason, *Biochem. Biophys. Res. Commun.*, **78**, 945 (1977).
47. M. Tanaka, M. Haniu, and K. T. Yasunobu, *Biochem. Biophys. Res. Commun.*, **76**, 1014 (1977).
48. M. M. Briggs and R. A. Capaldi, *Biochem. Biophys. Res. Commun.*, **80**, 553 (1978).
49. R. Bisson, H. Gutweniger, and A.*Azzi, *FEBS Lett.*, **92**, 219 (1978).
50. R. Bisson, A. Azzi, H. Gutweniger, R. Colonna, C. Montecucco, and A. Zanotti, *J. Biol. Chem.*, **253**, 1874 (1978).
51. M. E. Dockter, A. Steinemann, and G. Schatz, *J. Biol. Chem.*, **253**, 311 (1978).
52. M. Erecińska, *Biochem. Biophys. Res. Commun.*, **76**, 495 (1977).
53. R. Henderson, R. A. Capaldi, and J. S. Leigh, *J. Mol. Biol.*, **112**, 631 (1977).
54. T. G. Frey, S. H. P. Chan, and G. Schatz, *J. Biol. Chem.*, **253**, 4389 (1978).
55. G. D. Eytan, R. C. Carroll, G. Schatz, and E. Racker, *J. Biol. Chem.*, **250**, 8598 (1975).
56. U. Kunze and W. Junge, *FEBS Lett.*, **80**, 429 (1977).
57. M. Erecińska, D. F. Wilson, and J. K. Blasie, *Biochim. Biophys. Acta*, **501**, 53 (1978).
58. H. Blum, H. J. Harmon, J. S. Leigh, J. C. Salerno, and B. Chance, *Biochim. Biophys. Acta*, **502**, 1 (1978).
59. R. Malkin and B. G. Malmström, in *Advances in Enzymology*, Vol. 33, F. F. Nord, Ed., Wiley, New York, 1970, p. 177.
60. D. C. Wharton and A. Tzagoloff, *J. Biol. Chem.*, **239**, 2036 (1964).
61. M. Erecińska, B. Chance, and D. F. Wilson, *FEBS Lett.*, **16**, 284 (1971).
62. R. Weaver, J. H. van Drooge, G. van Ark, and B. F. van Gelder, *Biochim. Biophys. Acta*, **347**, 215 (1974).
63. L.-E. Andréasson, B. G. Malmström, C. Strömberg, and T. Vänngård, *FEBS Lett.*, **28**, 297 (1972).
64. T. Tsudzuki and D. F. Wilson, *Arch. Biochem. Biophys.*, **145**, 149 (1971).
65. B. Lanne and T. Vänngård, *Biochim. Biophys. Acta*, **501**, 449 (1978).
66. C. R. Hartzell, R. E. Hansen, and H. Beinert, *Proc. Natl. Acad. Sci. USA*, **70**, 2477 (1973).
67. H. Beinert, R. E. Hansen, and C. R. Hartzell, *Biochim. Biophys. Acta*, **423**, 339 (1976).
68. I. Salmeen, L. Rimai, and G. Babcock, *Biochemistry*, **17**, 800 (1978).
69. R. Aasa, S. P. J. Albracht, K.-E. Falk, B. Lanne, and T. Vänngård, *Biochim. Biophys. Acta*, **422**, 260 (1976).

70. D. F. Wilson, M. Erecińska, and C. S. Owen, *Arch. Biochem. Biophys.*, **175**, 160 (1976).
71. G. T. Babcock, L. E. Vickery, and G. Palmer, *J. Biol. Chem.*, **253**, 2400 (1978).
72. R. W. Shaw, R. E. Hansen, and H. Beinert, *Biochim. Biophys. Acta*, **504**, 187 (1978).
73. B. Lanne, B. G. Malmström, and T. Vänngard, *Biochim. Biophys. Acta*, **545**, 205 (1979).
74. F. R. N. Gurd, K.-E. Falk, B. G. Malmström, and T. Vänngard, *J. Biol. Chem.*, **242**, 5724 (1967).
75. C. R. Hartzell and H. Beinert, *Biochim. Biophys. Acta*, **423**, 323 (1976).
76. H. Beinert and R. W. Shaw, *Biochim. Biophys. Acta*, **462**, 121 (1977).
77. A. J. Thomson, T. Brittain, C. Greenwood, and J. Springall, *FEBS Lett.*, **67**, 94 (1976).
78. G. T. Babcock, L. E. Vickery, and G. Palmer, *J. Biol. Chem.*, **251**, 7907 (1976).
79. J. Peisach and W. E. Blumberg, *Arch. Biochem. Biophys.*, **165**, 691 (1974).
80. B. F. van Gelder and H. Beinert, *Biochim. Biophys. Acta*, **189**, 1 (1969).
81. B. Chance, C. Saronio, A. Waring, and J. S. Leigh, Jr., *Biochim. Biophys. Acta*, **503**, 37 (1978).
82. R. Brändén, J. Deinum, and M. Coleman, *FEBS Lett.*, **89**, 180 (1978).
83. K.-E. Falk, T. Vänngard, and J. Ångström, *FEBS Lett.*, **75**, 23 (1977).
84. M. F. Tweedle, L. J. Wilson, L. García-Iñiguez, G. T. Babcock, and G. Palmer, *J. Biol. Chem.*, **253**, 8065 (1978).
85. R. H. Tiesjema, A. O. Muijsers, and B. F. van Gelder, *Biochim. Biophys. Acta*, **305**, 19 (1973).
86. A. Cornish-Bowden and D. E. Koshland, Jr., *J. Mol. Biol.*, **95**, 201 (1975).
87. L.-E. Andéasson, *Eur. J. Biochem.*, **53**, 591 (1975).
88. B. G. Malmström, in *Symmetry and Function of Biological Systems at the Macromolecular Level*, A. Engström and B. Strandberg, Eds., Wiley, New York, 1969, p. 153.
89. Q. H. Gibson, C. Greenwood, D. C. Wharton, and G. Palmer, *J. Biol. Chem.*, **240**, 888 (1965).
90. Q. H. Gibson and C. Greenwood, *J. Biol. Chem.*, **240**, 2694 (1965).
91. E. Antonini, M. Brunori, C. Greenwood, B. G. Malmström, and G. C. Rotilio, *Eur. J. Biochem.*, **23**, 396 (1971).
92. S. Rosén, R. Brändén, T. Vänngård, and B. G. Malmström, *FEBS Lett.*, **74**, 25 (1977).
93. E. Antonini, M. Brunori, A. Colosimo, C. Greenwood, and M. T. Wilson, *Proc. Natl. Acad. Sci. USA*, **74**, 3128 (1977).
94. R. W. Shaw, R. E. Hansen, and H. Beinert, *J. Biol. Chem.*, **253**, 6637 (1978).
95. S. Rosén, *Biochim. Biophys. Acta*, **503**, 389 (1978).
96. B. Chance, C. Saronio, and J. S. Leigh, Jr., *J. Biol. Chem.*, **250**, 9226 (1975).
97. B. Chance and J. S. Leigh, Jr., *Proc. Natl. Acad. Sci. USA*, **74**, 4777 (1977).
98. B. Chance, C. Saronio, J. S. Leigh, Jr., W. J. Ingledew, and T. E. King, *Biochem. J.*, **171**, 787 (1978).
99. G. M. Clore and E. M. Chance, *Biochem. J.*, **173**, 799 (1978).
100. P. George, in *Oxidases and Related Redox Systems*, T. E. King, H. S. Mason, and M. Morrison, Eds., Wiley, New York, 1965, p. 3.

101. B. G. Malmström, L.-E. Andréasson, and B. Reinhammar, in *The Enzymes*, Vol. 12, Part B, P. D. Boyer, Ed., Academic Press, New York, 1975, p. 507.
102. T. Keevil and H. S. Mason, in *Methods in Enzymology*, Vol. 52, S. P. Colowick and N. O. Kaplan, Eds., Academic Press, New York, 1978, p. 3.
103. B. G. Malmström, *Biochem. J.*, **117,** 15P (1970).
104. R. Aasa, R. Brändén, J. Deinum, B. G. Malmström, B. Reinhammar, and T. Vänngård, *FEBS Lett.*, **61,** 115 (1976).
105. R. Aasa, R. Brändén, J. Deinum, B. G. Malmström, B. Reinhammar, and T. Vänngård, *Biochem. Biophys. Res. Commun.*, **70,** 1204 (1976).

CHAPTER **6**
Superoxide, Superoxide Dismutases and Oxygen Toxicity

JAMES A. FEE

Biophysics Research Division and
Department of Biological Chemistry,
The University of Michigan,
Ann Arbor, Michigan

CONTENTS

1 INTRODUCTION, 211
2 FORMATION OF SUPEROXIDE IN BIOLOGICAL SYSTEMS, 212
3 SUMMARY OF CHEMICAL PROPERTIES OF AQUEOUS SUPEROXIDE, 212

   3.1 Superoxide as a Reductant, 213
   3.2 Superoxide as an Oxidant, 213
   3.3 Superoxide Reactivity Toward Organic Substances in Water, 215

4 SUPEROXIDE DISMUTASES, 215
5 SUPEROXIDE TOXICITY, 218
6 FENTON CHEMISTRY, 220
7 EXTRAPOLATION FROM IN VITRO TO IN VIVO CONDITIONS, 223
8 BIOLOGICAL DISTRIBUTION AND THE FUNCTION OF SUPEROXIDE DISMUTASES, 227

   ACKNOWLEDGMENTS, 230

   REFERENCES, 230

1 INTRODUCTION

The anion formed by the one-electron reduction of oxygen is called superoxide, and it has been the subject of a great deal of research since the early 1930s (see references 1–4 for review of these early developments) and was first termed superoxide by Neumann in 1934 (5). The chemical and physical properties of superoxide have been the subject of several reviews (1–4, 6–8), and its potential importance in radiobiological experiments has been a point of scientific discussion for many years (10).

While the dominant reaction of superoxide in aqueous solution is its dismutation,

$$2O_2^- + 2H^+ \rightarrow H_2O_2 + O_2 \tag{1}$$

its ability to reduce a variety of chromophoric substances and the inhibition of the reduction process by superoxide dismutase have served as the bases of several assays for the presence of superoxide in systems reacting with oxygen (10). The first superoxide dismutase to be discovered was the Zn/Cu protein (11), which had previously been named erythrocuprein; this protein has become a valuable tool in the hands of enzymologists interested in the mechanisms of autoxidations.

The presence of this catalytic activity in living systems stirred anew questions concerning the role of superoxide in oxygen metabolism. McCord et al. (11) presented the hypothesis that oxygen toxicity[1] was mediated by superoxide. This was predicated on three points of conjecture: (a) Superoxide is formed in all cells utilizing molecular oxygen. (b) Superoxide or a product derived therefrom is toxic to cells. (c) Superoxide dismutases are present in all aerotolerant organisms for the purpose of minimizing the concentration of superoxide and thus providing protection against oxygen toxicity. As is considered in detail here, the facts strongly support the first part of this general idea, but the supporting evidence for the remaining parts of the theory is of a circumstantial nature.

A problem for the hypothesis that superoxide mediates oxygen toxicity is the demonstrable chemical *unreactivity* of O_2^- in aqueous medium. While this statement may appear to contradict the reader's previous experience, it has long been known that superoxide ion in aqueous solution is chemically benign toward most organic substances. Indeed extensive review (6) of the chemical literature has revealed no reaction between

---

There is no satisfactory molecular definition of oxygen toxicity. A reasonable working definition however is any degradation of a biological system which can be attributed to the presence of oxygen.

aqueous superoxide and a naturally occurring organic substance which could reasonably be considered deleterious to the cellular millieu. Nevertheless the biochemical literature is replete with demonstrations that O_2^- is involved in a variety of chemical processes which are clearly destructive of cellular material and which lead to cell death or severe modification of cellular activity. The question of why O_2^- displays unusual reactivity in in vitro biochemical systems but not in well-defined chemical experiments is one for which a satisfactory answer must be obtained before we can begin to assess the role of O_2^- in in vivo manifestations of oxygen toxicity. The purposes of this chapter are to resolve the above dilemma by offering a unifying explanation of in vitro manifestations of superoxide toxicity and to discuss this and other aspects of the field of superoxide and superoxide dismutases which are relevant to the potential role of superoxide in oxygen toxicity.

2 FORMATION OF SUPEROXIDE IN BIOLOGICAL SYSTEMS

There is no question that superoxide is formed by biological systems. Various protein systems, cellular extracts, isolated cell organelles, and even whole cells form superoxide to varying extents. Some of these are listed in Table 1. Estimates of the percentage of O_2^- formed during oxygen consumption range from 17% in extracts of *Streptococcus faecalis* (12) to <1% in submitochondrial particles (13). However there are good reasons to believe that the intactness of the subcellular structures is important in determining how much O_2^- is formed. In general the steady-state levels of O_2^- are very low, with estimates ranging from 10^{-7} to 10^{-10} M.

3 SUMMARY OF CHEMICAL PROPERTIES OF AQUEOUS SUPEROXIDE[2]

The superoxide anion free radical is strongly solvated by water molecules, an interaction which dominates its chemical reactivity. Kebarle and coworkers (73, 74) have shown that even in the gas phase O_2^- binds up to three water molecules with $\Delta H \simeq -17$ kcal/mole. A reasonable estimate of the heat of solution of O_2^- is -100 kcal/mole (75), which is similar to the -112 kcal/mole value of F^-. The degree and strength of interaction with water molecules are dominating influences on the chemical prop-

---

The reader is referred to references 6–8 for reviews of superoxide chemistry and an introduction to the extensive literature associated with this species.

erties of F^- and O_2^- (6, 8). Thus both anions are very powerful nucleophiles when dissolved in a weakly solvating liquid such as DMSO but appear to be poor nucleophiles in aqueous solution. Solvation also affects the affinity of these anions for protons. In aqueous media both are weak acids; $pK_{O_2^-} = 4.8$ and $pK_{F^-} = 3.5$. Thus O_2^- in aqueous solution is a poor nucleophile.

The reduction potentials for the one-electron reduction of oxygen have been discussed in reference 6. Superoxide can act as both a reductant ($E^{0\prime} = -0.16$ V)³ and an oxidant ($E^{0\prime} = +0.87$ V), and these features are included in the dismutation reaction.

3.1 Superoxide as a Reductant

The indicated potential suggests that O_2^- is a weak reductant, but its participation in a variety of chemical reactions demonstrates a facile propensity to transfer the electron to oxidizing molecules. Most of the reactions of O_2^- in aqueous solutions are simple one-electron reductions which often occur with great rapidity. Thus the reduction of Cu^{2+}, tetranitromethane, and Zn^{2+}/Cu^{2+} superoxide dismutase, for examples, occur at rates near diffusion control, and several other reductive reactions occur at lesser rates down to those where dismutation itself becomes a major competing process.

3.2 Superoxide as an Oxidant

There are only a few well documented examples of O_2^- acting as an oxidant of an organic molecule. Ascorbic acid is oxidized by O_2^- with $k = 3 \times 10^5$ at pH 7.3 (78), NADH bound to lactate dehydrogenase is oxidized to NAD , and the process is chain propagated in the presence of O_2 (79). α-Tocopherol (80) and Tiron (81) also appear to be oxidized by O_2^-. Reports that O_2^- can oxidize thiols are not interpreted as reliable, and experiments in our laboratory indicate no interference of thiols in the decay of O_2^- when adequate metal ion scavengers are included in the medium. Thus, while the reduction potential of $E^{0\prime} = +0.87$ V implies a strong oxidizing capability, the chemical evidence shows that O_2^- only

---

The potentials for half-reactions involving O_2 have traditionally used $p_{O_2} = 1$ atm for the standard state of O_2 (76). However most biochemical reactions occur on a time scale that does not allow equilibration with the gas phase, and a standard state of 1 M O_2 is thus reasonable. Note that this raises the potential of the $O_2 + e \rightleftharpoons O_2^-$ half-reaction from -0.33 (6) to -0.16, a value more in keeping with the ability of O_2 to be reduced by various substances (77).

Table 1 Representative Systems which Form Detectable[a] Amounts of Superoxide in the Presence of Oxygen

System	Method of Detection and References
Enzymes and Proteins	
Xanthine oxidase	E[c] (14, 15), C (10, 16)
Peroxidases	E (17), A (18), H (19)
Diamine oxidase	C (20)
Flavoproteins	C (21, 22)
Iron–sulfur proteins	E (23, 17), A (24)
P-450	CL (25), SC (26), I (27)
Hemoglobin	C (28, 32), A (29), MD (30, 31)
Rubredoxin	A (33)
Small Molecules	
Flavins	E (34), A (35)
Quinones	A (35), PR (36); C, MD (37, 38)
Bipyridilium herbicides	C, A (39), PR (40)
Phenazine methosulfate	N (41)
Tetrahydropteridines	N (42), MD (43)
Phenylhydrazine	MD (44, 45)
Dialuric acid	C (46–48)
Thiols	A (49)
Miscellaneous	
Sonication	C (50)
Electrolysis	C (51)
Melanin autoxidation	ST (52)
Protoporphyrin and light	ST (53)
Hydroxylamine autoxidation	MD (54)
Various redox dyes	NBT (55)
Cell Organelles	
Chloroplasts	R (56), ST (57), MD (58)
Arum maculatum SMP[d]	A (59)
Plant mitochondria and SMP[e]	A (13)
Mammalian mitochondria and SMP	A (60–63)
Microsomes[f]	A (64–66), NBT (67)[g]
Nuclei	A (69)
Cells	
Polymorphonuclear leukocytes	C (70, 71), R (72)

[a] The lower limit of superoxide detection is near 0.1 nmolar. Many of the experiments cited approach this limit.

[b] Only dehydrogenases appear to form O_2^- while oxidases and hydroxylases do not react with O_2 to form detectable levels of O_2^-.

[c] Methods of O_2^- detection and other notation: E, EPR: C, cytochrome c reduction; A, adrenochrome formation; H, inhibition of hydroxylation; CL, chem-

oxidizes organic molecules which are quite good reducing agents and which can donate positive charge as part of the oxidation reaction.

The kinetic "barrier" to one-electron oxidation has been proposed to lie in the fact that O_2^{2-} is highly energetic and therefore not readily formed (6). Thus a molecule which can only donate an electron is not expected to be oxidized by O_2^-, whereas a molecule capable of donating H·, that is, having a very acidic proton, may be oxidized by O_2^-.

Low-valency transition metal ions might also be expected to act as efficient reductants of O_2^-, and indeed Zn^{2+}/Cu^{1+} superoxide dismutase, Fe^{2+} superoxide dismutase, and several other metal ion complexes (6, 82, 83) transfer an electron to O_2^-, presumably to form a labile peroxo complex, while Fe^{2+}-EDTA reacts to form a relatively stable peroxo complex (84).

3.3 Superoxide Reactivity Toward Organic Substances in Water

Many attempts have been made to observe direct reactions between organic substances of biological importance and O_2^-. The available *negative* observations are listed in Table 2. These facts emphasize the chemical docility of aqueous superoxide and, when contrasted with the observations of many biochemists that O_2^- plays some role in a variety of degradative cellular processes, constitute the dilemma which has plagued this area of investigation almost since its inception.

In summary, superoxide can be quite reactive as a reductant in aqueous solution, but with the exception of metal ions it is generally a poor oxidant.

4 SUPEROXIDE DISMUTASES

Superoxide dismutases are metalloproteins which catalyze reaction (1). There are three distinct types of dismutases as indicated by the essential

---

iluminesence; SC, reduction of succinylated cytochrome c; I, inhibition of an O_2^--requiring enzymatic reaction; MD, methanistic deduction; PR, pulsed radiolysis; ST, spin trapping; R, review article; NBT, reduction of nitroblue tetrazolium.
[d] Only ~1.4% of the oxygen consumed by submitochondrial particles of this species proceeds via superoxide production.
[e] Greater than 95% oxygen reduction proceeds to water by way of the cyanide-sensitive and hydroxamic acid-sensitive pathways.
[f] There is some debate regarding the actual electron donor to O_2, but cytochrome P-450 appears to be the favored candidate.
[g] The extremely high ratios of O_2^- produced to oxygen consumed (~6) may be due to an artifact of the NBT assay recently described (68).

Table 2 Substances with Which Aqueous Superoxide Has Been Shown Not to React[a]

Substance	Condition or Type of Experiment	Reference
φx-174 DNA	Radiolysis/centrifugation	85
Nuclear bases	Gamma irradiation/production analysis	86
Poliovirus	KO_2 dissolution/virulence	87
Cholesterol	Product analysis	88
Formate, fumarate, α-ketogluterate, pyruvate, oxalate, imidazole, Tris, EDTA	Combined pulsed radiolysis and stopped-flow/H_2O_2 and O_2 analyses	89[b]
H_2O_2	KO_2 dissolution/O_2 analysis	90
	Stopped-flow kinetics	91
	H_2O_2 inhibition of NBT reduction by O_2^-	92
	Pulsed radiolysis	93
	Search for aromatic hydroxylation by O_2^- and H_2O_2	94, 95
Linolenic acid (dispersion)	Stopped-flow kinetics	Fee and McClune, unpublished
ROOR[c]		96
ROOH[c]		97
Hyaluronic acid	Xanthine/xanthine oxidase/O_2	B. Halliwell, personal communication

[a] Representative only.
[b] Several other related compounds were also found not to react with O_2^-.
[c] In acetonitrile.

metal cofactor; some of the properties of these proteins are presented in Table 3.

The Cu/Zn protein has been most widely studied (see reference 82 for a detailed review of the relationship between structural aspects of the metal center and catalysis of superoxide dismutation). In brief, elegant studies involving the rapid formation of O_2^- by pulsed radiolysis have shown that catalysis occurs by a cyclic redox pattern indicated by reactions (2) and (3) (98, 99):

$$O_2^- + P\text{—}Cu(II) \longrightarrow O_2 + P\text{—}Cu(I) \qquad (2)$$

$$O_2^- + P\text{—}Cu(I) \xrightarrow{H^+} H_2O_2 + P\text{—}Cu(II) \qquad (3)$$

Both oxidation and reduction steps occur at $\sim 2 \times 10^9$ m^{-1} sec^{-1} and are effectively limited by the rate at which O_2^- can diffuse to the Cu (98).

Much less is known about the iron- and manganese-containing proteins. Kinetic studies (100, 101) have revealed complexities which have been variously attributed to saturation of an anion binding site and/or to the formation of inactive states (83, 102).

Catalysis of superoxide dismutation is not limited to protein-bound metals. As shown in Table 4, the aquo complexes of Cu^{2+} and Mn^{2+} and a variety of small complexes of Cu^{2+} and Fe^{2+} are catalysts of reaction (1). Indeed Cu^{2+}_{aq} is approximately four times more effective than the Cu of the Zn/Cu superoxide dismutase!

What are the structural features required for a metal ion to possess superoxide dismutase activity? Previous discussions (82, 83) of this type of catalysis have pointed out the necessity for only three: (a) the metal ion must have at least one coordination position available for binding O_2^- in two adjacent valence states; (b) the redox potential of the metal should lie between ~ -0.1 and $+0.8$ V; and (c) the metal ion must be able to alternate between the two valence states more rapidly than the occurrence of spontaneous dismutation. For example,

$$Cu^{2+} + O_2^- \rightleftarrows Cu^{2+}\text{---}O_2^- \rightarrow Cu^+ + O_2$$

$$Cu^{1+} + O_2^- \rightleftarrows Cu^+\text{---}O_2^- \xrightarrow{H^+} Cu^{1+} + H_2O_2$$

While the coincidence of these three properties at the metal binding centers of the superoxide dismutases accounts for their dismutase activity, the fact that many other metal complexes also combine these properties and have this activity raises the possibility that the dismutase activity of

Table 3 General Properties of Superoxide Dismutases

Essential Cofactor	Molecular Weight	Number[a] of Subunits	Metals/ Subunit[b]	Source
Cu[c]	32,000	2	1 Zn, 1 Cu	eukaryotic cells
Fe	38,000	2	1 Fe	prokaryotic cells
Mn	40,000/80,000	2/4	1 Mn	pro- and eukaryotic cells

[a] Subunits appear to be identical.

[b] Idealized, fractions of a metal ion per subunit are generally found in purified preparations. It is not known whether this is due to incomplete occupancy of the metal binding site or to loss during purification.

[c] The function of Zn in this protein is not known, but it is not necessary for superoxide dismutase activity.

Table 4 Metal Ion Catalysts of Superoxide Dismutation

System	Relative Efficiency[a]	Reference
Bovine superoxide dismutase	1	10, 103
Cu^{2+}_{aq}	~4	104
Cu-Amino acids	~0.001–0.1	105–107
Cu-☐-Penicillamine	~0.01	108
Cu-Salicylate	0.001	109
Apo-BSD + Cu^{2+}	~1	110
Cytochrome oxidase	0.01–0.03	111
Ceruloplasmin	~0.0003	112[b]
Bacterial iron superoxide dismutase	1	113
Fe-EDTA	~0.001	83, 114
Fe-Porphyrin[c]	~0.001	115
Hemoglobin	0.001	116
Bacterial manganese superoxide dismutase	1	117
Mn-EDTA	0	118
Mn-Quinolinol[d]	+	119
Mn^{2+} in phosphate buffer	+	120

[a] Proteins are given a relative value of unity, although there appear to be small differences in absolute activity and remaining values are very approximate. A plus sign indicates some activity.
[b] Ceruloplasmin appears to scavenge O_2^- by a mechanism other than dismutation in a process not inhibited by cyanide.
[c] Tetrakis(4-N-methylpyridyl)porphine iron.
[d] In aprotic solvent.

the proteins called superoxide dismutases results from a fortuitous coincidence of these structural features at their metal binding sites.

5 SUPEROXIDE TOXICITY[4]

What are the manifestations of superoxide toxicity? Superoxide has been implicated in many in vitro phenomena: nonspecific hydroxylation of organic substances, degradation of DNA, lipid peroxidation, depolymerization of polysaccharide structures, and certain types of cell killing (Table

> The vigorous discussions concerning the role of superoxide in radiobiological systems presented in reference 9 are of considerable historical importance. Already in 1958 it was quite clear that O_2^- was known not to be of major importance in the damage of biological systems by ionizing radiation and to be generally benign toward organic substances.

Table 5 In vitro Degradative Processes in Which Superoxide Has Been Shown to Participate[a]

Process	Reference
DNA Degradation/depolymerization	121–124
Lipid peroxidation	125–132
Depolymerization of polysaccharides	133, 134[b]
Hydroxylation of aromatic substances	135–141
In vitro cell killing (bacteria and myoblasts)	142–146
Ethylene formation from methional[c]	147–149

[a] Representative. The participation of superoxide has generally been ascertained by an inhibition of the overall process by superoxide dismutase.
[b] Contains no experimental data supporting superoxide involvement.
[c] Including compounds related to methional.

5). These phenomena clearly contrast with the chemistry of superoxide described in the previous section. How can superoxide which has been shown by direct chemical observation to be unreactive toward the participating organic substances be essential in these myriad degradative processes?

The usual experimental indication of superoxide involvement has been at least a partial inhibition of the process under investigation by added superoxide dismutase. However almost without exception catalase also acts as an inhibitor. Beauchamp and Fridovich (147) were the first to recognize the interdependence of O_2^- and H_2O_2 in the production of an oxidizing species capable of initiating and perpetuating the complex degradative processes mentioned above. They proposed that O_2^- effected a reductive cleavage of H_2O_2 to form the extremely reactive free hydroxyl radical. Reaction (4),

$$O_2^- + H_2O_2 + H^+ \rightarrow OH\cdot + H_2O_2 + O_2 \qquad (4)$$

was suggested in 1934 by Haber and Weiss (150) to account for the quantitative aspects of peroxide decomposition in the presence of Fe salts. However it was known already in 1947 (151) that this reaction does not occur, a fact which has been confirmed repeatedly (see references 6 and 152 for summaries).[5]

It has been pointed out however that in certain instances reaction (4)

The proposal (127) that $^1\Delta O_2$ is also a product of reaction (4) seems most unlikely.

might be catalyzed by metal ions. We propose here that essentially all in vitro manifestations of superoxide toxicity can be explained by the following sequence of reactions:

$$M^{n+} + O_2^- \rightarrow M^{(n-1)+} + O_2 \qquad (5)$$

$$M^{(n-1)+} + H_2O_2 \xrightarrow{H^+} MO \quad \text{or} \quad (OH^\cdot + M^{n+}) + H_2O \qquad (6)$$

$$\left. \begin{array}{c} MO \\ \text{or} \\ OH^\cdot \end{array} \right\} + RH \rightarrow \left. \begin{array}{c} R^\cdot \\ \text{or} \\ ROH \end{array} \right\} \rightarrow \text{Degradation products} \qquad (7)$$

where M^{n+} is a metal such as Cu^{2+} or Fe^{3+} and MO is a metal-oxy compound having properties similar to the hydroxyl radical, $OH^\cdot$. The sum of reactions (5) and (6) is equal to reaction (4), which in the recent parlance of biochemists is called the Haber–Weiss reaction. However reaction (6) has been known among chemists for many years as the "Fenton reaction," and a more apt descriptive phrase for reactions (5) through (7) would be *superoxide-mediated Fenton chemistry*. While metal ions have long been known to be important in autoxidations, and this process has been mentioned in the context of specific cases by several authors (141, 152–154), the generalization of these reactions and the implications of extrapolating from the in vitro to the in vivo situation have not heretofore been considered (see editorial comments on page 320 in reference 82).

6 FENTON CHEMISTRY

Fenton chemistry is the generation of powerful oxidizing species from the reduction of peroxide by metal ions. Although discussed elsewhere in this volume, a brief review of Fenton chemistry seems appropriate at this point. It is clear from the known redox potentials (6, 76) that a very large decrease in free energy occurs upon reduction of H_2O_2 to H_2O ($E^{0'} = 1.35$ V) while very little decrease occurs ($E^{0'} = 0.38$ V) for formation of the hydroxyl radical from H_2O_2 by partial reduction. Thus most of the free-energy drop on converting H_2O_2 to H_2O resides in the reduction of $OH^\cdot$ to H_2O ($E^{0'} = 2.33$ V).[6] The powerful oxidizing capability and

---

There is still some uncertainty in the literature concerning the redox potential of the couples containing hydroxyl radical (155). The value quoted here is taken from George (76), however; Heckner and Landsberg (156) give an experimentally measured value of 1.4 ± 0.1 V.

FENTON CHEMISTRY

extreme kinetic reactivity of the hydroxyl radical toward organic substances are well documented (17).

The Fenton reaction, (6), is a potential source of the hydroxyl radical. However the real product of the reaction is still a hotly debated subject. One school argues that free hydroxyl radical is formed (157), while another argues for the production of a metal–oxy compound (158). Both substances however are considered to be powerful oxidants capable of H abstraction and insertion reactions which are surely able to initiate the degradative processes described in Table 5. The mechanism of this reaction is not important to this discussion.

Before proceeding to the individual processes in Table 5, let us look at the essential features of Fenton chemistry. In 1893 Fenton (159) reported that treatment of D-tartaric acid treated with H_2O_2 and Fe^{2+} yielded, among other products, cis-dihydroxymaleic acid (160). At the same time Ruff (161) discovered that α-hydroxycarboxylic acids were decarboxylated to aldehydes and CO_2 by a similar mixture. These observations resulted in a large body of work concerning the action of mixtures of H_2O_2 and transition metal salts on organic substances and of the chemical reactions resulting in the decomposition of H_2O_2 by metal salts.

The overall reaction schemes which now seem to be widely agreed upon are, using iron as an example,

Peroxide Decomposition

$$Fe^{2+} + H_2O_2 \xrightarrow{H+} \text{``OH''} + Fe^{3+} + H_2O \tag{6}$$

$$\text{``OH''} + H_2O_2 \longrightarrow O_2^- + H_2O + H^+ \tag{8}$$

$$Fe^{3+} + O_2^- \longrightarrow Fe^{2+} + O_2 \tag{9}$$

$$Fe^{3+} + H_2O_2 \longrightarrow Fe^{2+} + 2H^+ + O_2^- \tag{10}$$

$$2O_2^- + 2H^+ \longrightarrow H_2O_2 + O_2 \tag{1}$$

Alteration of Scheme by Organic Substances

$$Fe^{2+} + H_2O_2 \xrightarrow{H+} \text{``OH''} + Fe^{3+} + H_2O \tag{6}$$

$$\text{``OH''} + RH \longrightarrow \text{``R''} + H_2O \tag{7}$$

$$\text{``OH''} + H_2O_2 \longrightarrow O_2^- + H_2O + H^+ \tag{8}$$

$$Fe^{3+} + O_2^- \longrightarrow Fe^{2+} + O_2 \tag{9}$$

$$Fe^{3+} + H_2O_2 \longrightarrow Fe^{2+} + 2H^+ + O_2^- \tag{10}$$

The important fact to note here is that the solution must contain a reducing agent in order to recycle the metal ion after reaction (6). In the absence of other one-electron reductants, superoxide formed in reaction (8) has long been considered an important source of electrons for the oxidized metal. Peroxide itself, reaction (10), at high concentrations and pH may fulfill this role, but this is unlikely at physiological pH.

Other systems have been developed around this basic chemistry which provide for the formation of strong oxidants in dilute solutions of H_2O_2. These were developed by Udenfriend, Ullrich, and Hamilton (Table 6). The common features of all these systems is the presence of H_2O_2 (or a source thereof via autoxidation), reducing agent, and trace metals.

How can these facts be related to in vitro superoxide toxicity, that is, to the degradative processes listed in Table 5? A survey of relevant results published prior to the discovery of superoxide dismutase and of the early assertions that O_2^- was responsible for these degradative phenomena reveals them to have the same features as those systems specifically designed to support Fenton chemistry. This commonality is emphasized by the data presented in Table 7, which reveals that nonspecific hydroxylation of aromatic substances, ethylene production from methional, lipid peroxidation, DNA degradation, and polysaccharide depolymerization are all dependent on peroxide, metal ions, and an appropriate one-electron reducing agent. The assertion that superoxide serves exclusively as a reductant of the trace metal ions, reaction (5), in the systems listed in Table 5 has not been widely tested, but there are enough examples where

Table 6 Characteristics of Systems Producing a Strong Oxidizing Agent as a Result of H_2O_2 Reduction

System	Components	General Usage or Observations	Reference
Fenton's	Fe^{2+}, H_2O_2 ($\pm$ EDTA)	Hydroxylation, free-radical reactions	157–160, 145, 162, 163, 167
Ruff's	Fe^{2+}, H_2O_2	Decarboxylation of α-hydroxy acids	161
Udenfriend's	Fe^{2+}, Fe^{3+}, Cu^+, or Cu^{2+}/ascorbic acid/ O_2 or H_2O_2/EDTA (excess)	Hydroxylation of aromatic substances	164, 165, 167
Hamilton's	H_2O_2, catechol, Fe^{3+} or Fe^{2+}	Hydroxylation of anisole	166, 167
Ullrich's	Sn^{2+}/HPO_4^-/O_2	Hydroxylation	168, 169

the addition of a supplemental reductant obviates inhibition by superoxide dismutase to place considerable credence on this idea. These cases are summarized in Table 8.

These considerations support the idea that all the in vitro degradative processes in which superoxide has been shown to participate require the presence of trace metal ions, H_2O_2, and/or ROOH[7] and that the role of superoxide is primarily to reduce the trace elements so the Fenton reaction can occur. In this context the in vitro biochemical manifestations of superoxide toxicity are understandable in terms of the known chemical properties of superoxide ion in aqueous solution. Recent work of Aust (198) and Bors (199) and their collaborators confirm the essential nature of trace elements in in vitro degradative processes.

7 EXTRAPOLATION FROM IN VITRO TO IN VIVO CONDITIONS

What are the implications of extrapolating from the conclusions of in vitro experimentation to in vivo conditions? Since the superoxide formed within a cell will undoubtedly conform to the chemistry described above, one cannot expect a degradation of cell materials by any direct reaction of O_2^-. This leaves us with the question of whether superoxide mediated Fenton chemistry, reactions (5) through (7), makes an important contribution to the destructive forces imposed on a cell by the presence of molecular oxygen. The scheme presented in Figure 1 indicates how Fenton chemistry would be expected to occur in a cell including various "protective or avoidance and repair" mechanisms.

In the scheme, ΣM indicates all the metals in the cell that require reduction in order to participate in the subsequent reaction with peroxide. The relative velocities of precesses $1+$ and $2+$ defines the contribution of O_2^- to the total "Fenton toxicity" imposed on the cell by nonheme metals,[8] and it follows from the above arguments that this is the total contribution of O_2^- to oxygen toxicity.

The scheme is not meant to imply that Fenton chemistry is responsible for all manifestations of oxygen toxicity or to show all reactions which may involve oxygen. It is known that enzyme systems are sensitive to direct reactions with oxygen (203) as well as to a number of other pro-

---

The recent report (200) from Pryor's group indicating that O_2^- reduces ROOH directly is incorrect; it is an excellent example of the pervasiveness of adventitious metal ions in in vitro biochemical systems.

Porphyrin-associated iron, heme, does not appear to require a reducing agent to produce the strong oxidant analogous to the hydroxyl radical (201, 202).

Table 7 Fenton Conditions for Degradative Processes in Which Superoxide Participation has Been Observed

System	Conditions and Requirements	Reference
Ethylene production from "methional"[a]	$H_2O_2 + Fe^{2+}$	170
	O_2 + ascorbate and microsomes	171
	Sulfite, phenols and Mn^{2+}	172, 173
	O_2^- and H_2O_2 + "trace metals"[b]	147
DNA degradation, depolymerization or loss of biological activity	Effect of ascorbic acid, Cu^{2+}, O_2, and H_2O_2 on pneumococcal transforming substance	174[c]
	Effect of H_2O_2, KO_2, ascorbic acid, Fe-EDTA, and Cu-EDTA on poliomyelitis virus	87
	Effect of ascorbic acid, O_2, H_2O_2, Fe^{2+}, and a variety of other substances of similar nature	175–177[d]
	Quinones/quinols/cysteine, metal ions, and H_2O_2	123, 178–182[e]
Degradation of polysaccharides such as hyaluronic acid, alginic acids, and other polymeric structures	Effects of ascorbic acid, a variety of reducing agents, metal ions, chelators, and H_2O_2 on depolymerization	177, 183–191[f]
Peroxidation of lipids in a variety of biological structures under in vitro conditions	Metal ions, reducing agents, H_2O_2, and oxygen	192–197[g]
Hydroxylation of aromatics	See Table 6	
Cell killing	Treatment with riboflavin, light, and oxygen or various strongly reducing materials in the presence of normal culture components	47, 142, 143

[a] Here "methional" indicates similar substances such as methionine.
[b] "Trace metals" are apparently necessary for this process and are assumed here to be present even though not explicitly considered by the authors to be involved.
[c] McCarty (174) recognized the similarity of the destructive action of ascorbic acid on transforming substance with its effects on bacterial toxins, viruses, and enzymes described by previous workers.

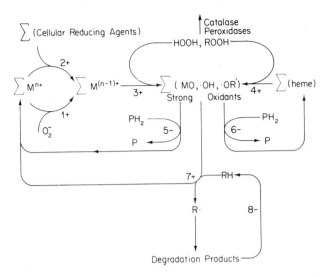

Figure 1. Scheme outlining possible contributions of Fenton chemistry to oxygen toxicity within a cell. M is a metal ion in the indicated relative valence state, MO is a metal–oxy compound having chemical properties similar to the hydroxyl radical, ·OH. PH_2 represents a substance which can be oxidized by the strong oxidants forming a product, P, which is not harmful to the cell. RH represents a molecule which when oxidized proceeds to degradation products deleterious to the cell. Σ indicates all available components able to participate in the given reaction. Numbered processes followed by (+) indicate movement toward degradation by Fenton-type chemistry; those followed by (−) suggest avoidance or repair of degradation.

cesses, but the quantitative contribution of these compared with Fenton-type reactions has not been assessed. However, since living organisms have harnessed the oxidizing potential of the hydroxyl radical by using enzymes such as catalase and peroxidase, it is reasonable that unwanted Fenton chemistry would be suppressed by specific sequestering of Fe and

[d] The authors of references 175–177 are clearly cognizant of the Fenton chemistry.
[e] While the authors of several of these papers do not explicitly involve metal ions in their proposed mechanism of degradation, it is clear such ions are ubiquitous in purified preparations of nucleic acids (183, and references therein).
[f] Pigman and co-workers (187–189) have termed the depolymerization of these substances the oxidative–reductive depolymerization (ORD) reaction, and its analogy to Fenton chemistry is obvious.
[g] This topic has been so widely studied that only a few general or leading references will be given which emphasize the essentiality of metal ions, reducing agents, and O_2 or peroxide.

Table 8 Ability of Added Reducing Agents to Substitute for Superoxide in Fenton-Type Systems with Loss of Inhibition by Superoxide Dismutase

System	Observation	Percent Inhibition by Superoxide Dismutase	Reference
Microsomal lipids/$FeCl_2$/ Fe-EDTA/O_2	Malondialdehyde formation		
+ Xanthine/xanthine oxidase		85	208
+ 0.2 mM ascorbic acid		2	
Brain lipids/metal ions/O_2	Fluorescent conjugate of peroxidized lipid		
+ 100 micromolar Fe^{2+} or Fe^{3+}		~43	209
+ 50 micromolar Cu^+ or Cu^{2+}		~60	
+ 1 mM ascorbate, no added metals[a]		0	
DNA/bleomycin/Fe/O_2	Malondialdehyde formation[b]	[c]	210, 211
Hyaluronic acid/xanthine oxidase/xanthine/O_2	Decrease in viscosity[d]		

[a] No overt attempt was made to minimize or sequester contaminating trace metals in this experiment, but it is certain they were present since extensive lipid peroxidation occurred.

[b] Acid-soluble radioactive fragments released from radiolabeled DNA are used as indicators of degradation as well as products which react with α-thiobarbituric acid to form colored products.

[c] The work described in references 210 and 211 establishes that the essential requirements for bleomycin-induced DNA degradation are Fe, O_2, and a reducing agent. A mixture of xanthine, xanthine oxidase, and O_2 can support Fe-bleomycin degradation and is inhibited by superoxide dismutase. However when other reducing agents are added such as thiols, there is virtually no inhibition by added superoxide dismutase. The authors propose a scheme whereby O_2^- can serve as a reductant of the Fe^{3+}-bleomycin complex which is also reduced by other reducing agents. These papers describe an excellent example of how the Fenton chemistry can be harnessed for a useful biological purpose.

[d] Metal ions are absolutely required for depolymerization, and added ascorbate obviates inhibition by superoxide dismutase (B. Halliwell, personal communication). The metal ion requirement is consistent with the observations of previous workers.

Table 9 Concentrations of Ascorbic Acid and Glutathione in Various Tissues[a]

Tissue	[GSH] $\times 10^3\,M$	[Ascorbate] $\times 10^3\,M$
Chloroplasts	3.5 (212)	2.5 (213)
Blue-green alga	>10 (214)[b]	>10 (214)[b]
Red blood cell	3 (215)	
Plasma		0.05 (216)
Lens	~20 (216)	~2 (216)

[a] Representative.
[b] Assuming a cellular volume of 2×10^{-15} liter.

Cu and by minimization of peroxide levels such that pathway 3+ is obviated. The little that is known of the reaction of reduced copper proteins and nonheme iron proteins with peroxides suggests that process 3+ can occur but is not rapid and generally requires high concentration of H_2O_2 (141, 204–207).[9] The intracellular concentrations of metal ions capable of contributing to cell damage is not known. However, because of the finite association constants of metal ions with cellular proteins, there will certainly be some free Cu and Fe ions in a cell, and this returns us to the questions of the relative velocities of pathways 1+ and 2+. Since cells generally contain high concentrations of reducing agents, Table 9, such as glutathione or cysteine and ascorbic acid, all of which are able to reduce the trace metals (217–219), it is reasonable to assume that the *velocity* of process 2+ will be much larger than that of pathway 1+, making the contribution of O_2^- to Fenton toxicity correspondingly small. Thus, unless there is a very specific, yet unknown destruction of essential cellular material by superoxide, the above considerations cast serious doubt on the hypothesis that superoxide is an important mediator of oxygen toxicity.

8 BIOLOGICAL DISTRIBUTION AND THE FUNCTION OF SUPEROXIDE DISMUTASES

McCord, Keele, and Fridovich (11) surveyed 26 microorganisms having widely different tolerance of oxygen for the presence of superoxide dis-

---

[9] The reader is referred to a series of papers by Oshino, Chance, and collaborators for elegant experiments concerning the role of H_2O_2 in hybaric oxygen toxicity (240).

mutase activity. Eight strict or facultative aerobes contained 1.4 to 7 units superoxide dismutase activity per mg protein, 10 strict anaerobes contained no dismutase activity, and eight aerotolerant anaerobes (microphiles) contained 0 to 1.6 units per mg protein. Thus there appeared to be a good correlation between dismutase activity and aerotolerance. Fridovich and his co-workers subsequently performed a number of experiments revealing induction of the manganese superoxide dismutase in bacteria by oxygen (220) and also by dyes which react with oxygen to form superoxide (221), and an adaptive resistance to bactericidal effects of hyperbaric oxygen which correlates well with increased levels of the manganese superoxide dismutase (222).

However these observations offer no *direct* evidence for the involvement of superoxide in in vivo manifestations of oxygen toxicity. The conclusions are somewhat clouded by the recent appearance of conflicting results and interpretations (223–225). Moreover we have recently shown that in addition to changes in manganese superoxide dismutase, methylviologen causes many proteins to increase, some as much as 30-fold, while many decrease by approximately a factor of 4.[10] These results demonstrate a global adaptation of the cell to the presence of the redox dye, and they make it impossible to assign to a single protein the greater resistance of the adapted cells to bacteriostatic and bactericidal redox dyes (221).

Circumstantial evidence is now beginning to accumulate which cannot readily be accommodated by the original hypothesis, and some of these observations are discussed in the following paragraphs.

In 1975 Hewitt and Morris (226) analyzed several anaerobic prokaryotes and found qualitative evidence for superoxide dismutase. They quantified the activity in two photosynthetic anaerobes, two sulfate reducers, and in 10 of 12 clostridia including two of the strains tested earlier (11). Quantitative measures of dismutase activity indicated considerable variability. However *Chlorobium thiosulfatophilum* (a photosynthetic anaerobe) had ~35% the activity of aerobically grown *Escherichia coli*, while the sulfate reducers contained from 1.5 to 6% and *Clostridium perfringens* had ~40% the level of *E. coli*. Gregory et al. (227) have recently surveyed 28 strains of the genus *Bacteroides* for superoxide dismutase activity and found activities ranging from 0.1 to 3.2 units/mg, compared to 1.2 units/mg for *E. coli* grown under anaerobic conditions. Generally lower levels and greater variability were found in clostridia, anaerobic cocci, and several

---

Unpublished work done in collaboration with A. Lees, P. Bloch, and F. C. Neidhardt (cf. also reference 225 for a further discussion of the complexities of this type of experiment).

other types of anaerobic bacteria. Tally et al. (228) examined 14 aerotolerant bacteria, four bacteria of intermediate oxygen sensitivity, and four which were extremely sensitive to oxygen. These authors found a very rough (if significant) correlation between oxygen tolerance and level of dismutase activity, and they proposed that oxygen tolerance may be an important factor in virulence of pathogenic anaerobes. However the data of Gregory et al. (227) run counter to this suggestion. There have been other reports of superoxide dismutase activity in anaerobic organisms (229–232.)

Iron-containing superoxide dismutase has been isolated and purified to homogeneity from *Desulfovibrio desulfuricans* (230), *Chromatium vinosum* (233), and *Chlorobium thiosulfatophilum* (234). The molecular, spectral, and catalytic properties of these proteins are closely analogous to the anaerobic superoxide dismutase of facultative *E. coli* (113).

The above observations made in several different laboratories definitely establish that the iron-containing superoxide dismutase is present in a wide variety of strict anaerobes at variable concentrations, which can be similar to those present in aerobic organisms. While several arguments have been advanced to rationalize the presence of superoxide dismutases in anaerobes, none is particularly convincing (228, 240), and the hypothesis of McCord et al. (11) must now be modified to fit these new facts. A minimal alteration was offered by Gregory (227): "This observation does not disprove the argument of McCord and Fridovich pertaining to obligate anaerobiosis but does argue that further definition of oxygen sensitivity and tolerance of anaerobes is needed." The question can be fairly asked as to whether superoxide dismutation is the true biological function of the metalloproteins presently called superoxide dismutases; indeed alternative biological functions for these proteins should be seriously considered.

One observation which suggests an alternate function is that certain aerobic cells produce inordinately high levels of superoxide dismutase. Thus manganese superoxide dismutase constitutes ~10% of the soluble protein of *Mycobacterium lepramurium* (235), and it is the dominant protein in the cytoplasm. This condition is more typical of proteins involved in a metabolic conversion essential to energy production than to a protective mechanism, although it is possible that the particular growth conditions used created a metabolic irregularity. Another example is the presence of millimolar levels of the Zn/Cu superoxide dismutase in seeds and seedlings. This protein thus constitutes 1.6 to 2.4% of the water-soluble protein in seedlings of corn, peas, and oats (236, 237) and somewhat lower levels in the corresponding seeds. Interestingly the embryo of corn, oat, and pea seeds contains approximately 10 times as much Zn/Cu dismutase as does the surrounding storage tissue. This suggests that the Zn/Cu protein in seeds is present for a reason other than protection against oxygen-free radicals. A likely explanation is that the Zn/Cu protein may act as a readily available storage and/or buffer source of Zn and Cu for use in newly synthesized metalloproteins during and after germination. A similar role in controlling Cu and Zn concentrations in eukaryotic cells is also possible (238).

ACKNOWLEDGMENTS

The author is indebted to Joan Valentine, Gregory McClune, and John Groves for many helpful discussions of the ideas presented above. These investigations were supported by USPHS Grant GM 21519.

REFERENCES

1. B. H. J. Bielski and J. M. Gebicki, in *Advances in Radiation Chemistry*, Vol. 2, M. Burton and J. L. Magee, Eds., Wiley–Interscience, New York, 1970, pp. 177–279.
2. G. Czapski, *Ann. Rev. Phys. Chem.*, **22**, 171 (1971).
3. N. G. Vannerberg, *Prog. Inorg. Chem.* **4**, 125 (1962).
4. I. V. Volnov, *Peroxides, Superoxides, and Ozonides of Alkali and Alkaline Earth Metals*, Plenum Press, New York, 1966.
5. E. W. Neumann, *J. Chem. Phys.*, **2**, 31 (1934).
6. J. A. Fee and J. S. Valentine, in *Superoxide and Superoxide Dismutases*, A. M. Michelson, J. M. McCord, and I. Fridovich, Eds., Academic Press, London, 1977, pp. 19–60.
7. E. Lee-Ruff, *Chem. Soc. Rev.*, **6**, 195 (1977).
8. J. S. Valentine, in *Oxygen: Biochemical and Clinical Aspects*, W. S. Caughey, Ed., Academic Press, in press.
9. R. Latarjet, Ed., *Organic Peroxides in Biology*, Pergamon Press, London, 1958.
10. J. M. McCord and I. Fridovich, *J. Biol. Chem.*, **244**, 6049 (1969).
11. J. M. McCord, B. B. Keele, Jr., and I. Fridovich, *Proc. Natl. Acad. Sci. USA*, **68**, 1024 (1971).
12. L. Britton, D. P. Malinowski, and I. Fridovich, *J. Bacteriol.* **134**, 299 (1978).
13. P. R. Rich and W. D. Bonner, Jr., *Arch. Biochem. Biophys.*, **188**, 206 (1978).
14. R. C. Bray, P. F. Knowles, F. M. Pick, and J. F. Gibson, *Hoppe-Seyl. Z.*, (1968).
15. P. F. Knowles, J. F. Gibson, F. M. Pick, and R. C. Bray, *Biochem. J.*, **111**, 53 (1969).
16. I. Fridovich and P. Handler, *J. Biol. Chem.*, **237**, 916 (1962).
17. R. Nilsson, F. M. Pick, and R. C. Bray, *Biochim. Biophys. Acta*, **192**, 145 (1969).
18. G. Rotilio, G. Falcioni, E. Fioretti, and M. Brunori, *Biochem. J.*, **145**, 405 (1975).

REFERENCES

19. B. Halliwell and S. Ahluwalia, *Biochem. Soc. Trans.*, **4**, 73 (1976).
20. G. Rotilio, L. Calabrese, A. Finazzi-Agro, and B. Mondovi, *Biochem. Biophys. Acta*, **198**, 618 (1970).
21. V. Massey, S. Strickland, S. G. Mayhew, L. G. Howell, P. C. Engel, R. G. Matthews, M. Schuman, and P. A. Sullivan, *Biochem. Biophys. Res. Commun.*, **36**, 891 (1969).
22. H. J. Forman and J. A. Kennedy, *Biochem. Biophys. Res. Commun.*, **60**, 1044 (1974).
23. W. H. Orme-Johnson and H. Beinert, *Biochem. Biophys. Res. Commun.*, **36**, 905 (1969).
24. H. P. Misra and I. Fridovich, *J. Biol. Chem.*, **246**, 6886 (1971).
25. S. G. Sligar, J. D. Lipscomb, P. G. DeBrunner, and I. C. Gunsalus, *Biochem. Biophys. Res. Commun.*, **61**, 290 (1974).
26. H. Kuthan, H. Tsuji, H. Graf, V. Ullrich, J. Werringloer, and R. W. Estabrook, *FEBS Lett.*, **91**, 343 (1978).
27. H. W. Strobel and M. J. Coon, *J. Biol. Chem.*, **246**, 7826 (1971).
28. R. Wever, B. Oudega, and B. F. Van Gelder, *Biochim. Biophys. Acta*, **302**, 475 (1973).
29. H. P. Misra and I. Fridovich, *J. Biol. Chem.*, **247**, 6960 (1972).
30. W. J. Wallace, J. C. Maxwell, and W. S. Caughey, *Biochem. Biophys. Res. Commun.*, **57**, 1104 (1974).
31. W. J. Wallace and W. S. Caughey, *Biochem. Biophys. Res. Commun.*, **62**, 561 (1975).
32. L. S. Demma and J. M. Salhany, *J. Biol. Chem.*, **252**, 1226 (1977).
33. S. W. May, B. J. Abbott, and A. Felix, *Biochem. Biophys. Res. Commun.*, **54**, 1540 (1973).
34. D. Ballou, G. Palmer, and V. Massey, *Biochem. Biophys. Res. Commun.*, **36**, 898 (1969).
35. H. P. Misra and I. Fridovich, *J. Biol. Chem.*, **247**, 188 (1972).
36. K. B. Patel and R. L. Willson, *J. Chem. Soc. Faraday Trans.*, **1**, 814 (1973).
37. R. Heikkila and G. Cohen, *Science*, **172**, 1275 (1971).
38. R. E. Heikkila and G. Cohen, *Science*, **181**, 456 (1973).
39. T. C. Stancliffe and A. Pirie, *FEBS Lett.*, **17**, 297 (1971).
40. J. A. Farrington, M. Ebert, E. J. Land, and K. Fletcher, *Biochim. Biophys. Acta*, **314**, 372 (1973).
41. M. Nishikimi, N. A. Rao, and K. Yagi, *Biochem. Biophys. Res. Commun.*, **46**, 849 (1972).
42. M. Nishikimi, *Arch. Biochem. Biophys.*, **166**, 273 (1975).
43. R. E. Heikkila and G. Cohen, *Experientia*, **31**, 169 (1975).
44. H. P. Misra and I. Fridovich, *Biochemistry*, **15**, 681 (1976).
45. B. Goldberg, A. Stern, and J. Peisach, *J. Biol. Chem.*, **251**, 3045 (1976).
46. J. A. Fee, R. Bergamini, and R. G. Briggs, *Arch. Biochem. Biophys.*, **169**, 160 (1975).
47. L. R. DeChatelet, P. S. Shirley, P. R. Goodson, and C. E. McCall, *Antimicrob. Agents Chemother.*, **8**, 146 (1975).
48. G. Cohen, R. Heikkila, *J. Biol. Chem.*, **249**, 2447 (1974).
49. H. P. Misra, *J. Biol. Chem.*, **249**, 2151 (1974).
50. B. Lippitt, J. M. McCord, and I. Fridovich, *J. Biol. Chem.*, **247**, 4688 (1972).

51. H. J. Forman and I. Fridovich, *Science*, **175**, 339 (1972).
52. C. C. Felix, J. S. Hyde, T. Sarna, and R. C. Seally, *Biochem. Biophys. Res. Commun.* **84**, 335 (1978).
53. G. R. Buettner and L. W. Oberley, *FEBS Lett.*, **98**, 18 (1979).
54. Y. Kono, *Arch. Biochem. Biophys.*, **186**, 189 (1978).
55. S. Dupré, G. Federici, L. Santoro, M. R. Rossi Fanelli, and D. Carallini, *Mol. Cell. Biochem.*, **9**, 149 (1975).
56. K. Asada, M.-A. Takahashi, K. Tanaka, and Y. Nakano, in *Biochemical and Medical Aspects of Active Oxygen*, O. Hayaishi, and K. Asada, Eds., Japan Scientific Society Press, Tokyo, 1977, pp. 45–63.
57. J. R. Harbour and J. R. Bolton, *Biochem. Biophys. Res. Commun.*, **64**, 803 (1975).
58. B. L. Epel and J. Neumann, *Biochim. Biophys. Acta*, **325**, 520 (1972).
59. S. Huq and J. M. Palmer, *Plant Sci. Lett.*, **11**, 351 (1978).
60. G. Loschen, A. Azzi, C. Richter, and L. Flohe, *FEBS Lett.*, **42**, 68 (1974).
61. H. Nohl and D. Hegner, *Eur. J. Biochem.*, **82**, 563 (1978).
62. O. Dionisi, T. Galeotti, T. Terranova, and A. Azzi, *Biochim. Biophys. Acta*, **403**, 292 (1975).
63. E. Cadenas, A. Boveris, C. I. Ragan, and A. O. M. Stoppani, *Arch. Biochem. Biophys.*, **180**, 248 (1977).
64. S. D. Aust, D. L. Roerig, and T. C. Pederson, *Biochem. Biophys. Res. Commun.* **47**, 1133 (1972).
65. G. M. Bartoli, T. Galeotti, G. Palombini, G. Parisi, and A. Azzi, *Arch. Biochem. Biophys.*, **184**, 276 (1977).
66. D. H. Nelson and A. Ruhmann-Wennhold, *J. Clin. Invest.*, **56**, 1062 (1975).
67. C. Auclair, D. deProst, and J. Hakim, *Biochem. Pharm.*, **27**, 355 (1978).
68. C. Auclair, M. Torres, and J. Hakim, *FEBS Lett.*, **89**, 26 (1978).
69. G. M. Bartoli, J. Galeotti, and A. Azzi, *Biochim. Biophys. Acta*, **497**, 662 (1977).
70. B. M. Babior, R. Kipnes, and J. Curnette, *J. Clin. Invest.*, **52**, 741 (0000).
71. D. B. Drath and M. L. Karnovsky, *J. Exp. Med.*, **141**, 257 (1975).
72. B. M. Babior, *New England J. Med.*, **298**, 659 and 721 (1978).
73. P. Kebarle, M. Arshadi, and J. Scarborough, *J. Chem. Phys.*, **49**, 917 (1968).
74. M. Arshadi and P. Kebarle, *J. Phys. Chem.*, **74**, 1483 (1970).
75. R. Yamdagni, J. D. Payzant, and P. Kebarle, *Can. J. Chem.*, **51**, 2507 (1973).
76. P. George, in *Oxidases and Related Redox Systems*, Vol. I. T. E. King, H. S. Mason, and M. Morrison, Eds., Wiley, New York, 1965, pp. 3–36.
77. I. Yamazaki and L. H. Piette, *Biochim. Biophys. Acta*, **77**, 47 (1963).
78. M. Nishikimi, *Biochim. Biophys. Res. Commun.*, **63**, 463 (1975).
79. B. H. J. Bielski and P. C. Chan, *J. Biol. Chem.*, **251**, 2841 (1976).
80. M. Nishikimi and L. J. Machlin, *Arch. Biochem. Biophys.*, **170**, 684 (1975).
81. C. L. Greenstock and R. W. Miller, *Biochim. Biophys. Acta*, **396**, 11 (1975).
82. J. A. Fee, in *Superoxide and Superoxide Dismutases*, A. M. Michelson, J. M. McCord, and I. Fridovich, Eds., Academic Press, London, 1978, pp. 172–192.
83. J. A. Fee and G. J. McClune, in *Mechanisms of Oxidizing Enzymes*, T. P. Singer and R. N. Ondarza, Eds., Elsevier–North-Holland, Amsterdam, 1977, pp. 273–284.

REFERENCES

84. G. J. McClune, J. A. Fee, G. A. McClusky, and J. T. Groves, *J. Am. Chem. Soc.*, **99**, 5220 (1977).
85. J. Blok, in *Radiation Research*, G. Silini, Ed., North-Holland, Amsterdam, 1967, pp. 423–437.
86. J. Cadet and R. Teoule, *Photochem. Photobiol.*, **28**, 661 (1978).
87. Y. Pelouz, C. Nofre, A. Cier, and L. Colobert, *L'Inst. Pasteur Anal.*, **102**, 6 (1962).
88. L. L. Smith, M. J. Kulig, and J. I. Teng, *Chem. Phys. Lipids*, **20**, 221 (1977).
89. B. H. J. Bielski and H. W. Richter, *J. Am. Chem. Soc.* **99**, 3019 (1977).
90. P. George, *Disc. Faraday Soc.*, **2**, 196 and 219 (1947).
91. G. J. McClune and J. A. Fee, *FEBS Lett.*, **67**, 294 (1976).
92. B. Halliwell, *FEBS Lett.*, **72**, 8 (1976).
93. G. Czapski and Y. A. Ilan, *Photochem. Photobiol.*, **28**, 657 (1978).
94. J. A. Fee and P. G. Hildenbrand, *FEBS Lett.*, **39**, 79 (1974).
95. A. Rigo, R. Stevanato, A. Finazzi-Agro, and G. Rotilio, *FEBS Lett.*, **80**, 130 (1977).
96. J. W. Peters and C. S. Foote, *J. Am. Chem. Soc.*, **98**, 873 (1976).
97. M. J. Gibian and T. Ungermann, *J. Am. Chem. Soc.*, **101**, 1291 (1979).
98. E. M. Fielden, P. B. Roberts, R. C. Bray, D. J. Lowe, G. N. Mautner, G. Rotilio, and L. Calabrese, *Biochem. J.*, **139**, 49 (1974).
99. D. Klug-Roth, I. Fridovich, and J. Rabani, *J. Am. Chem. Soc.*, **95**, 2786 (1973).
100. F. Lavelle, M. E. McAdam, E. M. Fielden, K. Puget, and A. M. Michelson, *Biochem. J.*, **161**, 3 (1977).
101. M. E. McAdam, R. A. Fox, F. Lavelle, and E. M. Fielden, *Biochem. J.*, **165**, 71 (1977).
102. M. E. McAdam, F. Lavelle, F. A. Fox, and E. M. Fielden, *Biochem. J.*, **165**, 81 (1977).
103. E. M. Fielden, P. B. Roberts, R. C. Bray, D. J. Lowe, G. N. Mautner, G. Rotilio, and L. Calabrese, *Biochem. J.*, **139**, 49 (1974).
104. J. Rabani, D. Klug-Roth, and J. Lilie, *J. Phys. Chem.*, **77**, 1169 (1973).
105. D. Klug-Roth and J. Rabani, *J. Phys. Chem.*, **80**, 588 (1976).
106. R. Brigelius, R. Spottl, W. Bors, E. Lengfelder, M. Saran, and U. Weser, *FEBS Lett.*, **47**, 72 (1974).
107. M. Younes and U. Weser, *FEBS Lett.*, **61**, 209 (1976).
108. M. Younes and U. Weser, *Biochim. Biophys. Res. Commun.*, **78**, 1247 (1977).
109. L. R. DeAlvare, K. Goda, and T. Kimura, *Biochem. Biophys. Res. Commun.*, **69**, 687 (1976).
110. J. A. Fee, R. Natter, and G. S. T. Baker, *Biochim. Biophys. Acta*, **295**, 96 (1973).
111. K. A. Markossian, A. A. Poghassian, N. A. Paitian, and R. M. Nalbandyan, *Biochem. Biophys. Res. Commun.*, **81** 1336 (1978).
112. I. M. Goldstein, H. B. Kaplan, H. S. Edelson, and G. Weissman, *J. Biol. Chem.*, **254**, 4040 (1979).
113. T. O. Slkyhouse and J. A. Fee, *J. Biol. Chem.*, **251**, 5472 (1976).
114. B. Halliwell, *FEBS Lett.*, **56**, 34 (1976).
115. R. F. Pasternack and B. Halliwell, *J. Am. Chem. Soc.*, **101**, 1026 (1979).
116. C. C. Winterbourn, R. E. Hawkins, M. Brian, and R. W. Carrell, *J. Lab. Clin. Med.*, **85**, 337 (1975).
117. B. B. Keele, Jr., J. M. McCord, and I. Fridovich, *J. Biol. Chem.* **245**, 6176 (1970).

118. J. Stein, J. P. Fackler, G. J. McClune, and J. A. Fee, in preparation.
119. J. K. Howie and D. T. Sawyer, *J. Am. Chem. Soc.*, **98**, 6698 (1976).
120. K.-L. Chen and P. B. McCay, *Biochem. Biophys. Res. Commun.*, **48**, 1412 (1972).
121. J. J. Van Hemmen and W. J. A. Meuling, *Biochim. Biophys. Acta*, **402**, 133-141 (1975).
122. J. R. White, T. O. Vaughn, and W. S. Yeh, *Fed. Proc.*, Abst. #537 (1971).
123. R. J. Lorentzen and P. O. P. Tso, *Biochemistry*, **16**, 1467 (1977).
124. R. Cone, S. K. Hasan, J. W. Lown, and A. R. Morgan, *Can. J. Biochem.* **54**, 219 (1976).
125. J. A. Fee and H. D. Teitelbaum, *Biochem. Biophys. Res. Commun.*, **49**, 150 (1972).
126. R. E. Lynch and I. Fridovich, *J. Biol. Chem.*, **253**, 1838 (1978).
127. E. W. Kellog, III, and I. Fridovich, *J. Biol. Chem.*, **250**, 8812 (1975).
128. D. D. Tyler, *Biochim. Biophys. Acta*, **396**, 335 (1975).
129. T. C. Pederson and S. D. Aust, *Biochem. Biophys. Res. Commun.*, **48**, 789 (1972).
130. J. S. Bus, S. D. Aust, and J. E. Gibson, *Biochem. Biophys. Res. Commun.*, **58**, 749 (1974).
131. D. D. Tyler, *FEBS Lett.*, **51**, 180 (1975).
132. T. Noguchi and M. Nakano, *Biochem. Biophys. Acta*, **368**, 446 (1974).
133. J. M. McCord, *Science*, **185**, 529 (1974).
135. H. S. Isbell and H. L. Frusch, *Carbohyd. Res.*, **59**, C25 (1977).
136. S. Strickland and V. Massey, in *Oxidases and Related Redox System*, T. E. King, H. S. Mason, and M. Morrison Eds., University Park Press, 1973, pp. 189-194.
137. S. A. Goscin and I. Fridovich, *Arch. Biochem. Biophys.*, **153**, 778 (1972).
138. B. Halliwell, *Eur. J. Biochem.*, **55**, 355 (1975).
139. B. Halliwell and S. Ahluwalia, *Biochem. J.*, **153**, 513 (1976).
140. B. Halliwell, *Biochem. J.*, **163**, 441 (1977).
141. J. M. McCord and E. D. Day, Jr., *FEBS Lett.*, **86**, 139 (1978).
142. F. Lavelle, A. M. Michelson, and L. Dimitrijevic, *Biochem. Biophys. Res. Commun.*, **55**, 350 (1973).
143. A. M. Michelson and M. E. Buckingham, *Biochem. Biophys. Res. Commun.*, **58**, 1079-;086 (1974).
144. E. M. Gregory, F. J. Yost, and I. Fridovich, *J. Bacteriol.* **115**, 987 (1973).
145. L. R. DeChatelet, P. S. Shirley, P. R. Gordson, C. E. McCall, *Antimicrob. Agents Chemother.*, **8**, 146 (1975).
146. J. J. Van Hemmen and W. J. A. Meuling, *Arch. Biochem. Biophys.*, **182**, 743 (1977).
147. C. Beauchamp and I. Fridovich, *J. Biol. Chem.*, **245**, 4641 (1970).
148. S. J. Weiss, P. K. Rustagi, and A. F. LoBuglio, *J. Exp. Med.*, **147**, 316 (1978).
149. S. J. Weiss, J. Turk, and P. Needleman, *Blood*, in press.
150. F. Haber and J. Weiss, *Proc. Roy. Soc.*, **A147**, 332 (1934).
151. P. George, *Disc. Faraday Soc.*, **2**, 196 and 219 (1947).
152. B. Halliwell, *Biochem. J.*, **167**, 317 (1977).
153. K. L. Fong, P. B. McCay, J. L. Poyer, H. P. Misra, and B. B. Keele, *Chem. Biol. Interact.*, **15**, 77 (1976).
154. E. F. Elstner, M. Saran, W. Bors, and E. Lengfelder, *Eur. J. Biochem.*, **89**, 61 (1978).

REFERENCES

155. W. H. Koppenol, *Nature*, **262**, 420 (1976).
156. K. H. Heckner and R. Landsberg, *J. Phys. Chem.*, **222**, 63 (1965).
157. C. Walling, *Acc. Chem. Res.*, **8**, 125 (1975).
158. M. L. Kremer, *Israel J. Chem.*, **9**, 321 (1971).
159. H. J. H. Fenton, *J. Chem. Soc.*, **65**, 899 (1893).
160. H. J. H. Fenton, *J. Chem. Soc.*, **69**, 546 (1896).
161. O. Ruff, *Ber.*, **31**, 1573 (1898).
162. J. H. Merz and W. A. Water, *J. Chem. Soc.*, 2427 (1949).
163. G. Sosnosky and D. J. Rawlinson, in *Organic Peroxides*, D. Swern, Ed., Vol. II, Wiley-Interscience, New York, 1971, pp. 269-336.
164. S. Udenfriend, C. T. Clark, J. Axelrod, and B. B. Brodie, *J. Biol. Chem.*, **208**, 731 (1954).
165. B. B. Brodie, J. Axelrod, P. A. Shore, and S. Udenfriend, *J. Biol. Chem.*, **208**, 741 (1954).
166. G. A. Hamilton and J. P. Friedman, *J. Am. Chem. Soc.*, **85**, 1008 (1963).
167. R. O. C. Norman and J. R. Lindsay Smith, in *Oxidases and Related Redox Systems*, Vol. I, T. E. King, H. S. Mason, and M. Morrison, Eds. Wiley, New York, 1965, pp. 131-156.
168. V. Ullrich, D. Hey, J. Staudinger, H. Buch, and W. Rummel, *Biochem. Pharmacol.* **16**, 2237 (1967).
169. V. Ullrich and H. Staudinger, in *Microsomes and Drug Oxidations*, J. R. Gillette, A. H. Conney, G. J. Cosmides, R. W. Estabrook, J. R. Fouts, G. J. Mannerling (Eds.) Academic Press, London, 1969, pp. 199-223.
170. M. Lieberman, A. T. Kunishi, L. W. Mapson, and D. A. Wardale, *Biochem. J.*, **97**, 449 (1965).
171. M. Lieberman and P. Hochstein, *Science*, **152**, 213 (1966).
172. S. F. Yang, *Arch. Biochem. Biophys.*, **122**, 481 (1967).
173. S. F. Yang, *J. Biol. Chem.*, **244**, 4360 (1969).
174. M. McCarty, *J. Exp. Med.*, **81**, 501 (1945).
175. G. Limperos and W. A. Mosher, *Am. J. Roent. Rad. Ther.*, **63**, 581 (1950).
176. B. E. Conway and J. A. V. Butler, *J. Chem. Soc.*, 834 (1952).
177. D. L. Gilbert, R. Gerschman, J. Cohen, and W. Sherwood, *J. Am. Chem. Soc.*, **79**, 5677 (1957).
178. J. Morita, S. Morita, and T. Komano, *Biochim. Biophys. Acta*, **475**, 403-407 (1977).
179. H. S. Rosenkranz and S. Rosenkranz, *Arch. Biochem. Biophys.*, **146**, 483 (1971).
180. H. R. Massie, H. V. Samis, and M. B. Baird, *Biochim. Biophys. Acta*, **272**, 539 (1972).
181. R. Cone, S. K. Hasan, J. W. Lown, and A. R. Morgan, *Can. J. Biochem.*, **54**, 219 (1976).
182. H. L. White and J. R. White, *Biochim. Biophys. Acta*, **123**, 648 (1966).
183. V. A. Srinska, L. Messineo, R. L. R. Towns, and K. H. Pearson, *Int. J. Biochem.*, **9**, 637 (1978).
184. W. B. Robertson, M. W. Ropes, and W. Bauer, *Biochem. J.*, **35**, 903 (1941).
185. A. G. Ogston and T. F. Sherman, *Biochem. J.*, **72**, 301 (1959).
186. P. Alexander and M. Fox, *J. Polym. Sci.*, **33**, 493 (1958).

187. W. Pigman and S. Rizvi, *Biochem. Biophys. Res. Commun.*, **1**, 39 (1959).
188. W. Pigman, S. Rizci, and H. L. Holley, *Arthritis Rheum.*, **4**, 240 (1961).
189. G. Matsumara and W. Pigman, *Arch. Biochem. Biophys.*, **110**, 526 (1965).
190. W. Niedermeier, C. Dobson, and R. P. Laney, *Biochim. Biophys. Acta*, **141**, 366-373 (1967).
191. W. Niedermeier, R. P. Laney, and C. Dobson, *Biochim. Biophys. Acta*, **148**, 400 (1967).
192. K. U. Ingold in *Lipids and their Oxidation*, H. W. Schultz, E. A. Day, and R. O. Sinnhuber, Eds., Avi Press, Westport, Connecticut 1962, pp. 93–121.
193. E. D. Wills, *Biochim. Biophys. Acta*, **98**, 238 (1965).
194. E. D. Wills, *Biochem. J.*, **113**, 325 (1969).
195. F. E. Hunter, Jr., G. M. Gebicki, P. E. Haffstein, J. Weinstein, and A. Scott, *J. Biol. Chem.*, **238**, 828 (1963).
196. R. C. McKnight, F. E. Hunter, and W. H. Oehlert, *J. Biol. Chem.*, **240**, 3439 (1965).
197. H. F. Deutsch, B. E. Kline, and H. P. Rusch, *J. Biol. Chem.*, **141**, 529 (1947).
198. B. A. Svingen and S. D. Aust, in *Molecular Basis of Environmental Toxicity*, R. S. Bhatnagar, Ed., Ann Arbor Science Publishers, Ann Arbor, in press (1979).
199. W. Bors, C. Michel, and M. Saran, *Eur. J. Biochem.*, **95**, 621 (1979).
200. M. J. Thomas, K. S. Mehl, and W. A. Pryor, *Biochem. Biophys. Res. Commun.*, **83**, 927 (1978).
201. A. L. Tappel, in *Lipids and Their Oxidation* H. W. Schultz, E. A. Day, and R. O. Sinnhuber, Eds., Avi Press, Westport, Connecticut, 1962, pp. 122–138.
202. R. M. Kaschnitz and Y. Hatefi, *Arch. Biochem. Biophys.*, **171**, 292 (1975).
203. M. Itakura and E. W. Holmes, *J. Biol. Chem.*, **254**, 333 (1979).
204. A. Mazur, S. Green, and E. Shorr, *J. Biol. Chem.* **220**, 227 (1956).
205. W. L. Anderson and T. B. Tomasi, Jr., *Arch. Biochem. Biophys.*, **182**, 705 (1977).
206. R. C. Bray, S. A. Cockle, E. M. Fielden, P. B. Roberts, G. Rotilio, and L. Calabrese, *Biochem. J.*, **139**, 43 (1974).
207. E. K. Hodgson and I. Fridovich, *Biochemistry*, **14**, 5294 (1975).
208. T. C. Pederson and S. D. Aust, *Biochem. Biophys. Res. Commun.*, **52**, 1071 (1973).
209. J. M. C. Gutteridge, *Biochem. Biophys. Res. Commun.*, **77**, 379 (1977).
210. E. A. Sausville, J. Peisach, and S. B. Horwitz, *Biochemistry*, **17**, 2740 (1978).
211. E. A. Sausville, R. W. Stein, J. Peisach, and S. B. Horwitz, *Biochemistry*, **17**, 2746 (1978).
212. C. H. Foyer and B. Halliwell, *Plenta*, **133**, 21 (1976).
213. B. Gerhardt, **61**, 101 (1964).
214. S. R. Tözüm and J. R. Gallon, *J. Gen. Microbiol.*, **11**, 313 (1979).
215. E. Bernt and H. U. Bergmeyer, in *Methods of Enzymatic Analysis*, Vol 4, H. U. Bergmeyer, Ed., Academic Press, New York, 1974, pp. 1643–1647.
216. A. White, P. Handler, and E. L. Smith, *Principles of Biochemistry* McGraw-Hill, New York, 1973.
217. K. A. Skov and D. J. Vonderschmitt, *Bioinorg. Chem.*, **4**, 199 (1975).
218. P. L. Leussing and L. Newman, *J. Am. Chem. Soc.*, **78**, 552 (1956).
219. D. Cavallini, C. DeMarco, S. Dupre, and G. Rotilio, *Arch. Biochem. Biophys.*, **130**, 354 (1969).

REFERENCES

220. E. M. Gregory and I. Fridovich, *J. Bacteriol.* **114,** 543 (1973).
221. H. M. Hassan and I. Fridovich, *J. Biol. Chem.*, **253,** 8143 (1978).
222. E. M. Gregory and I. Fridovich, *J. Bacteriol.* **114,** 1193 (1973).
223. L. Britton and I. Fridovich, *J. Bacteriol.* **131,** 815 (1977).
224. R. S. Simons, P. S. Jackett, M. E. W. Carroll, and D. B. Lowrie, *Toxicol. Appl. Pharmacol.* **37,** 271 (1976).
225. J. A. Fee, A. C. Lees, P. L. Bloch, and F. C. Neidhardt, in *Oxygen: Biochemical and Clinical Aspects,* W. S. Caughey, Ed., Academic Press, in press.
226. J. Hewitt and J. G. Morris, *FEBS Lett.,* **50,** 315 (1975).
227. E. M. Gregory, W. E. C. Moore, and L. V. Holdeman, *Appl Environ. Microbiol.* **35,** 988 (1978).
228. F. P. Tally, B. R. Goldin, N. V. Jacobus, and S. L. Gorbach, *Infect. Immunity,* **16,** 20 (1977).
229. D. G. Lindmark and M. Müller, *J. Biol. Chem.,* **249** 4634 (1974).
230. E. C. Hatchikian and Y. A. Henry, *Biochemie,* **59,** 153 (1977).
231. J. Carlsson, J. Wrethen, and G. Beckman, *J. Clin. Microbiol.,* **6,** 280 (1977)
232. K. Yano and H. Nishie, *J. Gen. Appl. Microbiol.,* **24,** 333 (1978).
233. S. Kanematsu and K. Asada, *Arch. Biochem. Biophys.,* **185,** 473 (1978).
234. S. Kanematsu and K. Asada, *FEBS Lett.,* **91,** 94 (1978).
235. K. Ichihara, E. Kusunose, M. Kusunose, and T. Mori, *J. Biochem.,* **81,** 1427 (1977).
236. C. N. Giannopolitis and S. K. Ries, *Plant Physiol.,* **59,** 309 (1977).
237. C. N. Giannopolitis and S. K. Ries, *Plant Physiol.,* **59,** 315 (1977).
238. J. S. Valentine and S. J. Lipparal, submitted for publication.
239. N. Oshino and B. Chance, *Biochem. J., * **162,** 509 (1977).
240. J. Lumsden and D. O. Hall, *Nature,* **257,** 670 (1975).

Index

Acetamides, 140
Acid metmyoglobin, 98
  MCD spectrum of, 101
Adamantane, oxidation of, 140
Adenosine triphosphate, 12
  synthesis of, 183
Alkanes, oxidation by chromic acid, 134
  oxidation by ferrate (VI) salts, 140
  reactions of metal-oxo compounds with, 134-140
Alkenes, reactions of metal-peroxo compounds with, 128-133
Alkyl imidazole, as axial base, 35
Allosteric effects, in kinetic experiments, 200-201
Aminopeptidase, 6
Aromatics, hydrolylation of, superoxide participation, 224
Ascaris lumbricoides, 13, 59
  O_2 affinity of, 59
Ascorbic acid, concentration in tissues, 227
  oxidation by superoxide, 213
Atmungsferment, 184, 185

Bacillus macerans, 171
Bacterial cytochrome, 78
Bacteroides, superoxide dismutase activity, 228
Beef heart cytochrome oxidase, peptide subunits, 188
5,6-Benzoflavone-inducible P-450$_{LM4}$, 78
Benzoyl chloride, reaction of platinum-dioxygen complexes with, 133
Benzphetamine, 86
  binding to cytochrome P-450, 88
  determination by gel filtration, 88
N-Benzyl-1,4-dihydronicotinamide, as reducing agent, 140
Bioinorganic chemistry, synthetic analogs in, 3-5
Biological oxygen transfer, chemical models of, 140-146
Bisimidazole heme, 43
Bohr effect, 12
Bromelain, 89
3,5-di-*t*-Butylcatechol, reaction of oxygen with, 176
n-Butyllithium, 128

Camphor-hydroxylating enzyme, 77
Camphor hydroxylation, 77
Candida, O_2 affinity of, 59
Capped porphyrins, 20, 45, 47, 55
  stability of, 55, 56
Carbon-monoxide-binding pigment, 75
Carbonmonoxyhemoproteins, photo-dissociation of CO from, 36
Carboxypeptidase, 6
Catalase, 219
Catechols, 176, 178
  autoxidation, 176
  oxygen insertion into, 176
  as reducing agents, 140
Catechols, substituted, relative rates of reaction for, 178
Ceruloplasmin, 6
Chlorobium thiosulfatophilum, 228
  dismutase activity, 228

isolation of iron-containing superoxide dismutase from, 229
m-Chloroperoxybenzoic acid, reaction of $P450_{LM2}$ with, 116-117
Chlorophyll, 6
5-Chlorosalicylic acid, catabolism by extracts of *Bacillus brevis*, 166
cleavage by a dioxygenase, 165
Chlorotetraphenylporphinato iron(III), 156
Chromatium vinosum, isolation of iron-containing superoxide dismutase from, 229
Chromic acid, oxidation of alkanes by, 134
Chromyl acetate, oxidation of aldehydes by, 135
Chromyl chloride, reaction with *cis*-alkenes, 138
Circular dichroism, 98-99
Clostridium perfringens, dismutase activity, 228
Cobalt (II) heme, 28
Cobalt hemoglobin, O_2 affinity of, 59
Cobalt hemoproteins, O_2 affinities of, 58
Cobalt-*meso*-tetraphenylporphyrin(1,2-dimethylimidazole), 60
Cobalt myoglobin, O_2 affinity of, 59
Cobalt porphyrins, 28-33
  axial coordination chemistry of, 28, 29
  complexes of, 30
  O_2 complexes, 31-33, 57-59
  equilibrium constants for binding of axial bases to, 29
  picket fence, 51
  reaction with Lewis bases, 28
Cobalt (protoporphyrin IX dimethyl ester), 31
Cobalt protoporphyrin IX dimethyl ester-(1-methylimidazole), 59
Cobalt-substituted hemoglobin, 28
Cobalt-substituted myoglobin, 28
Cobalt (t-Bsalten) (1-benzimidazole) (O_2), crystal structure of, 30
Cobalt tetra(p-methoxyphenyl)porphyrin-(1-methylimidazole)
Cobalt tetratolylporphyrins, 59
Co-*meso*-α,α,α,α-tetra-*o*-pivalamidophenylporphyrin(1,2-dimethylimidazole), 60
Cooperativity, 28

mechanism of, 19
in oxygen binding, mechanism, 60, 62-66
Copper-catalyzed oxygenations, 146
Copper signal, 195, 196, 200
Copper/zinc protein, 216
Cumene hydroperoxide, 116
Cupric bromide, 137
Cyclobutyl tosylate, solvolysis of, 136-137
Cyclohexanol, 109, 147, 150
  hydroxylation of, 150
  oxidation of, 147
Cyclohexene, oxidation of, 153
Cysteine, concentration in tissues, 227
Cytochrome a, 185, 186, 193, 202
Cytochrome a_3, 186, 193, 194, 200, 201, 202
Cytochrome b_5, 86, 87
Cytochrome c oxidase, 9, 181-203
  biological distribution, 183
  catalytic reaction, 199-202
    allosteric effects, 200
    electron transfer, sequence of, 199-200
  composition, 187-188
    copper-to-heme ratio, 186, 188
    phospholipid content, 187-188
  isolation, chromatographic procedures, 187
  magnetic properties, 196-197
  oxidation-reduction properties, 197-199
    redox titrations based on optical measurements, 197-198
  prosthetic groups of, 185-187
  reactions of, 183-184
  reduced reoxidation by dioxygen, 200, 201
  reduction of, electron-transfer mechanism, 186-187
  role in oxidative phosphorylation, 183
  spectroscopic properties, 191-196
    EPR spectra, 192-196
    infrared spectra, 191-192
    visible spectra, 191-192
  structure, 185, 187-190
    membrane arrangement, 189-190
    peptide subunits, 188-189
Cytochrome P-420, 75
Cytochrome P-450, 73, 78-82
  as catalyst in monooxygenation reactions, 73
  catalytic cycle of, 104-116

INDEX

catalyzed reactions, 91
  positional selectivity, 91
  stereospecificity, 91
circular dichroism of, 98-99
interaction with reductase, 90
properties of, 79
spectroscopic behavior of, 91-103
  circular dichroism spectra, 98-99
  electronic spectroscopy, 92-96
  electron paramagnetic resonance spectroscopy, 96-97
  infrared spectroscopy, 102
  magnetic circular dichroism, 100-102
  Mössbauer spectroscopy, 97-98
  nuclear magnetic resonance, 102
  resonance Raman spectroscopy, 102, 103
structure of, 91-92
Cytochrome P-450$_{cam}$, 81, 82
Cytochrome P-450$_{LM2}$, absorption spectra of, 80, 86
  EPR spectra of, 82
Cytochrome P-450$_{LM4}$, 78, 79, 81, 86
  absorption spectra of, 80
  EPR spectra of, 82
Cytochrome P-450 reductase, 103
Cytochrome P-450-reductase-phospholipid system, 113
Cytochromes, 5, 6, 75
  rabbit, 79

Deoxyerythrocruorin, structure of, 51
Deoxy ferrous myoglobins, 7
Deoxyhemoglobin, 193
Deoxy iron porphyrinates, structures of, 52
Deoxymyoglobin, 193
Desulfovibrio desulfuricans, isolation of iron-containing superoxide dismutase from, 229
Detergent-solubilized reductase, 84
trans-2-Deuteriocyclohexanol, oxidation of, 154
Endo-deuterium, 110
Exo-Deuterium, 110
Diaminopurine, as reducing agent, 140
1,2-Dihydroxydihydronaphthalene, 141
cis-Dihydroxymaleic acid, 221
3,4-Dihydroxyphenyl propionate, as substrate, 172

Dioxygen, binding to heme proteins, 14
Dioxygenases, mechanism of action, 163-178
  reactions, catalysis of, 167
  structure of, 167-171
2,3-Diphosphoglycerate, 12
Dismutase activity, correlation between oxygen tolerance and, 229
DNA degradation, 222
  superoxide participation, 224

Effectors, competitive, 12-13
  quaternary, 12
  tertiary, 12
Electrochemical oxidation, of iron salts, 149
Electronic spectroscopy, 92-96
Electron paramagnetic resonance spectroscopy, 96-97
Epoxidation reactions, 133, 157
Erythrocruorins, definition of, 13
Erythrocuprein, 211
Escherichia coli, dismutase activity, 228
Ethylene, production from methional, 222
  superoxide participation, 224
Extended X-ray absorption fine structure spectroscopy (EXAFS), 54

Fe-*meso*-α, α, α, α-tetra-*o*-pivalamidophenylporphyrin (1-methylimidazole), 63, 64
Fe-*meso*-α,α,α,α-tetra-*o*-pivalamidophenylporphyrin (2-methylimidazole). EtOH, 60, 62-65
Fe-*meso*-tetraphenylporphyrin (2-methylimidazole), 60
Fe-*meso*-tetraphenylporphyrin (2-methylimidazole) C$_2$H$_5$OH, 17
[Fe(*meso*-tetraphenylporphyrin) (OH$_2$)$_2$]$^+$, 21
Fe porphine catalysts, 157
Fe porphyrinate-O$_2$ complexes, 25, 47-54
  physical properties of, 47-54
  electronic spectra, 47
  Mössbauer spectra, 48-49
  vibrational spectra, 47
  stability to irreversible oxidation, 54-55
  structural data, 51-54
Fe prophyrinates, cooperativity in O$_2$ binding to, 62-66

Fe porphyrin complexes, electronic
   spectra, 95
  magnetic circular dichroism of, 101
  Mössbauer data for, 98
Fe porphyrins, dioxygen binding to, 33-66,
   57
  kinetic studies, 35-39
  low-temperature stabilization, 33-35
  polymer-supported systems, 39-45
  protected pocket porphyrins, 45
  dioxygen complexes of, 27
Fe protoporphyrin IX, 95
Fe protoporphyrin IX dimethyl ester, 95, 97
Fe (II) deuteroporphyrin, 20
Fe (II) dioxygen complexes, stabilization of,
   45
Fe (II) *meso*-tetraphenylporphyrin (1-methyl-
   imidazole)$_2$, reaction with O$_2$, 34
Fe (II) *meso*-tetraphenylporphyrin (piperidine)$_2$, reaction with O$_2$ in dry
   CH$_2$Cl$_2$, 34
Fe (II) *meso*-tetraphenylporphyrin
   (pyridine)$_2$, reaction with O$_2$ in dry
   CH$_2$Cl$_2$, 34
Fe (II) octaethylporphyrin, 41
Fe (II) porphyrinate, 12
Fe (II) porphyrin-imidazole systems, CO
   binding, 66
  O$_2$ binding, 66
Fe (II) porphyrins, autooxidation, 27
  axial coordination of, 14-22
  equilibrium constants for binding of
   axial bases to, 17
  oxidation by O$_2$, 25-27
  oxygen complexes, stabilization of, 27
  preparation of, 16
  spin-state properties of, 16
  structure of, 16
Fe (II) protoporphyrinate IX complex, 7
Fe (II) protoporphyrin IX dimethyl ester,
   41
Fe (II) superoxide dismutase, 215
Fe (II) tetra(*m*-tolyl) porphyrin, 26
Fe (III) deuteroporphyrin IX dimethyl
   ester, 24
Fe (III) *meso*-tetraphenylporphyrin, 24
Fe (III) porphyrin complexes, EPR spectra
   of, 97
Fe (III) porphyrins, 22-24
  axial coordination chemistry of, 22-24
  equilibrium constants for aromatic
   nitrogen base binding to, 24
Fe (III) tetra(p-sulfonato)phenylporphyrin,
   24
Fenton chemistry, 220-223
  contributions to oxygen toxicity within a
   cell, 225
  degradative processes, with superoxide
   participation, conditions for, 224
Fenton's reagent, 146-147
Fenton-type systems, ability of reducing
   agents to substitute for superoxide in,
   226
Fermi coupling, 48
Ferrate (VI) salts, oxidation of alkanes by,
   140
Ferric perchlorate hydrate, photoreduction
   of, 149
Ferricyanide, 84, 85, 86
  oxidation of semiquinone by, 84
Ferritin, 6
Ferrous carbonyl cytochrome, 113
Ferrous ion-hydrogen peroxide mixtures,
   see Fenton's reagent
Ferrous ion-mercaptobenzoic acid oxygen
   system, 141
Ferrous mesoporphyrins, 20
Ferrous porphyrins, *see* Fe (II) porphyrins
Flavoproteins, 83

Gastrophilus intestinales, 13, 59
  O$_2$ affinity of, 59
Globins, 7
Glutathione, concentration in tissues, 227
Glyceryl-3-phosphorycholine, acyl deriva-
   tives, 86

Half-reactions, involving oxygen, 213
Hayaishi, Osamu, 165
Heme, 9, 11, 13
Heme *a*, 185
  chemical structure of, 186
Heme cofactor, 7
Heme/polymer systems, oxygenation of,
   39, 40
Heme proteins, dioxygen binding to, 14
Heme proteins, oxygen binding to, 1-66
  kinetics of, 36
Hemerythrin, 5, 6
Heme signals, 195

copper signal, 195
  in oxidized cytochrome oxidase, 193
Heme systems, polymer-bound, 44
Hemocyanin, 5, 6
Hemoglobin-O_2, 27
  autooxidation, 27
  stabilization of, 27
Hemoglobin, "R"-state, 60
  O_2 affinity, 60
  "T"-state, 60-62
  O_2 binding to, 60-62
Hemoglobins, 3-4, 6, 9-13, 98
  bioinorganic chemistry, 3-4
  definition, 13
  fetal, 13
  Mössbauer spectra, 48-49
  oxygen association, 10-12, 14, 59
    kinetics of, 36, 37
  oxygen transport by, 13
  structure of, 9-10
Hemoproteins, 9
  dioxygen binding in, 14
  high-affinity, O_2 affinities of, 59
  magnetic susceptibility measurements of, 50
  Mössbauer spectral data of, 50
  nonmammalian, 13-14
  oxygen affinities of, 37, 57, 58
  oxygen-carrying, 5-14
    structure and function of, 5-14
  P-450, 92
    spectral properties, 92-93
  "R"-type, thermodynamic values for O_2 binding to, 56, 57, 59
  structural data, 13, 51
  subunits, 13
  "T"-type, thermodynamic values for O_2 binding to, 56, 60-62
  visible spectra of, 48
Hill equation, 11
Hill plot, of O_2 binding to hemoglobin, 10
Histidine, 19, 20
Hoard-Perutz cooperativity mechanism, see Cooperativity, mechanism of
Homotropic interactions, 11
Horseradish peroxidase, 26
Hydrazobenzene, as reducing agent, 140-141
Hydrogen peroxide reduction, systems producing a strong oxidizing agent as a result of, 222

Hydrolases, 6
Hydroperoxidases, 5
Hydroperoxide-supported hydroxylations, 114-116
p-Hydroxybenzoate, 171
α-Hydroxy-γ-carboxy-cis,cis-muconic semialdehyde, cyclization of with ammonium ions, 170
α-Hydroxycarboxylic acids, decarboxylation of, 221
Hydroxylase, 6
Hydroxylation, of aromatic substances, 222
  with ferrous iron-hydrogen peroxide, 146-149
  with ferrous iron-peroxyacids, 150-155
7-Hydroxynorborane, hydroxylation of, 148

Imidazole ligands, attachment to polymeric supports, 42
Imidazole polymer/heme systems, reversible oxygenation in, 41
Imidazoles, 15, 19, 20, 24, 25, 95
Indoleamine-2,3-dioxygenase, 144
Indole derivatives, oxidative cleavage of, 143-144
Indophenol oxidase, 184
Intradiol mechanism, $^{18}O_2$ insertion by, 165
Intramolecular isotope effect, 110
Iodosylbenzene, as oxidant, 157, 158
Iron porphyrin complexes, see Fe porphyrin complexes
Iron porphyrins, see Fe porphyrins
Iron salts, electrochemical oxidation of, 149
Iron tetra(pivalamidophenyl) porphyrin, 95
Iron (II) tetraaza macrocycle, 32
Isomerases, 6
2-Isopropylimidazole, 61

Lavoisier, 165
Leghemoglobin, 13
  O_2 affinity of, 36, 37, 59
Lewis base, 28
Lipid peroxidation, 222
  superoxide participation, 224
Liver microsomal cytochrome P-450 system, 76, 78
  rate-limiting step in, 112

see also Cytochrome P-450$_{LM2}$; Cytochrome P-450$_{LM4}$
Liver microsomal enzyme system, 75, 86
  components, 86
  electron transfer reactions in, 77
  reconstituted substrate hydroxylation in, 87
Liver microsomal monooxygenation system, components of, 78-89
Liver microsomes, 82
2,4-Lutidinic acid, 170

Magnetic circular dichroism, 100-102
Manganese-oxo compounds, hydroxylation of aliphatic centers by, 138-139
Manganese superoxide dismutase, 228
Mercaptobenzoic acid, as reducing agent, 140
Meso-tetra($\alpha,\alpha,\alpha,\alpha$-o-aminophenyl)porphyrin, 46
Meso-tetra($\alpha,\alpha,\alpha,\alpha$-o-pivalamidophenyl)porphyrin, 46
Meso-tetra(o-aminophenyl)porphyrin atropisomers, 46
Meso-tri(α,α,α-o-pivalamidophenyl)-β-o-aminophenylporphyrin, 46
Metal-catalyzed oxygen insertion, mechanisms of, 125-158
Metal-containing biomolecules, classification of, 6
Metal-dioxygen complexes, 128
Metal ions, association constants with cellular proteins, 227
  structural features for superoxide dismutase activity, 217
Metalloporphyrins, 155
  axial coordination chemistry of, 14
  five-coordinate, geometry of, 18
  oxidations catalyzed by, 133-158
Metalloproteins, 3
Metal-oxo compounds, reactions with alkanes, 134-140
Metal-peroxo compounds, reactions with alkenes, 128-133
Metal-to-oxygen electron transfer, 127
Methemoglobin, 98
Methional, ethylene production from, 222, 224
  superoxide participation, 224
N-Methylimidazole, 95
(1-Methylimidazole)Co(meso-tetraphenyl-
(1-Methylimidazole)Co(meso-tetraphenylporphyrin
N-Methylimidazole ferrous carbonyl porphyrin, 101-102
(3-Methylpyridine)$_2$Co(octaethylporphyrin), 30
2-Methyltetrahydrofuran, 95
2-Methyl-1-vinylimidazole, copolymer of, with styrene, 41
Methylviologen, 228
Metmyoglobin, 9
Metmyoglobin hydroxide, 194
Microsomal peroxidase, 115
Mitochondria, 189
Molybdenum(meso-tetraphenylporphyrin)(O$_2$)$_2$, 21
Molybdenum (V) peroxo compounds, 130
  as epoxidizing reagents, 130
Monooxygenases, 5
5-Mono(p-acrylamidophenyl)-10,15,20-triphenyl-porphine), 43
Mössbauer spectral data, of hemoproteins, 50
Mössbauer spectroscopy, 97-98
cis,cis-Muconic acid, 165, 177
cis,cis-Muconic acid monomethyl ester, 146
Mycobacterium lepramurium, manganese superoxide dismutase in, 229
Myoglobin, 3-4, 6, 7-9
  bioinorganic chemistry of, 3
  biological function of, 9
  components, 7
  definition of, 13
  ligand binding properties, 8-9
  occurrence, 7
  oxygen binding to, 7, 12, 59
    kinetics of, 36, 37
    Mössbauer spectra, 48-49
    tertiary structure, 7-8
Myoglobin-O$_2$, 27
  autooxidation, 27
  stabilization of, 27

NADH, 75, 77
NADH- and NADPH-dependent pathways, interactions of, 87
NADH-cytochrome b$_5$ reductase, 87
NADPH, 75, 76, 85-89, 114-115
NADPH-cytochrome, 103
NADPH-cytochrome P-450 reductase, 76, 83-86, 115
  interaction with P-450$_{LM}$, 88, 90

purification from detergent-solubilized rat liver microsomes, 83
Naphthalene, oxidation of, 141, 142
$[(NH_3)_{10}Co_2O_2]^{4+}$, 30
Nitrogenase, 6
(NO)Co(*meso*-tetraphenylporphyrin), 30
Norbornane, hydroxylation of, catalyzed by P-450$_{LM2}$, 111
2-Norborneol, 110
Norcarane, oxidation of, 153
Nucleophiles, 138, 213

Octaethyl porphyrin, complexes of, 24
Oxenoid mechanism, 108, 109
Oxidase, 6
Oxidases, terminal, 5
Oxidative phosphorylation, 5, 13
Oxido-reductases, 6
μ-Oxo dimers, 45
μ-Oxoporphyrins, 23
Oxyerythrocruorin, structure of, 53
Oxygen, activation of, 111-113, 166-167
 toxicity of, 211
 definition, 211n
Oxygenases, 75, 165-166
 definition, 165
Oxygenations, copper-catalyzed, 146
Oxygen binding, 1-66
 cooperativity in, 62-66
 to heme proteins, 1-66
 to "R"-type hemoproteins, 55-60
 stability to irreversible oxidation, 54-55
 thermodynamics of, 54-66
Oxygen insertion reactions, 134
 metal-catalyzed, 125-158
Oxygen rebound, 110, 111, 112
Oxygen transfer, biological, chemical models of, 140-146
Oxyhemoproteins, crystal structure, 51
Oxy iron porphyrinates, structures of, 52

P-450, *see* Cytochrome P-450
P450$_{LM2}$, reaction with *m*-chloroperoxybenzoic acid, 116-117
Palladium (II)-peroxo complexes, 128
Peroxide decomposition, 221-222
Peroxochromium, 128
μ-Peroxo dimers, 45
Peroxo metallacycles, 131-132
Peroxouranium (IV) oxide, reactions with olefins, 132
Peroxymolybdenum, 128
Perutz mechanism, of hemoglobin cooperativity, 11
Phenobarbital-inducible P-450$_{LM2}$, 78
2-Phenyl-1-vinylimidazole, copolymer of, with styrene, 41
Phosphatases, 6
Phosphates, organic, 13
Phosphatidylcholine, 86
 binding to cytochrome P-450, 89
Phospholipids, 76, 86-87
Picket fence porphyrins, 17, 31, 32, 45, 47, 49, 51, 53, 56, 59, 61, 62, 66, 96
 cooperativity of complexes, 55, 56
 stability of, 55, 56
 synthesis of, 45
(Piperidine)$_2$Fe(*meso*-tetraphenylporphyrin), x-ray crystal structure of, 21
Pivaloyl chloride, 46
Platinum-dioxygen complexes, reactions with benzoyl chloride, 133
Platinum(II)-peroxo complexes, 128
Pollutants, detoxification by hydroxylation, 165
Poly-*l*-lysine/heme, reaction of O$_2$ with, 41
Polyphosphazene polymer, imidazole-functionalized, 41
Polysaccharides, depolymerization of, 222
 superoxide participation, 224
Polystyrene, 39, 40
Poly(4-vinylpyridine) /heme, reaction of O$_2$ with, 41
Porphyrin(2-methylimidazole)Fe(*meso*-$\alpha,\alpha,\alpha,\alpha$,tetra-o-pivalamidophenyl-porphyrin, 17
Porphyrins, 15
 with covalently attached bases, 34
 covalent attachment to amine-containing polymers, 43
 imidazole base, oxygen binding, 39
 pyridine base, oxygen binding, 39
 see also Capped porphyrins; Picket fence porphyrins; Tailbase porphyrins
Prokaryotes, anaerobic, 228
Protected pocket porphyrins, 45-66
 synthesis, 45-47
Proteins, respiratory, 5
Protocatechuic acid, cleavage of, 170
 by dioxygenases, 167

Protocatechuic acid, 2,3-oxygenase, 171, 175, 177
Protocatechuic 3,4-dioxygenase, 176
  dissociation constants for, 174
  EPR parameters for, 175
Protocatechuic acid, 3,4-oxygenase, 167-170, 171, 172
  composition, 167-169
  Mössbauer spectra of, 173
  reaction mechanism for, 177
  structure, 168-169
Protocatechuic acid 4,5-oxygenase, 170-171, 175, 177, 178
Protoporphyrin, 11
Protoporphyrin IX dimethyl ester, 29
Pseudomonas aeruginosa, 167, 168
Pseudomonas fluorescens, 167
Pseudomonas putida, 75, 77, 168
Pseudomonas testosteroni, 170
Putidaredoxin, 77
Pyridines, 15
  as axial bases, 35
Pyrocatechase, 165, 168, 170
Pyromellitic acid, 20
Pyrroporphyrins, 20

Rabbit liver microsomal cytochrome P-450, substrate specificity of forms of, 91
Rabbit liver microsomal enzymes, 78
Rat liver reductase, 84
Redox-active metal centers, 185
Reductase, interaction with cytochrome P-450, 90
  detergent-solubilized, 84
Resonance Raman spectroscopy, 47
Respiratory proteins, 184
Rhodium (I) complexes, 131
Rubredoxin, 172

Bis(Salicylidene)ethylenediamine(salen), cobalt (II) complexes of, 143
Semiquinone, air-stable, oxidation by ferricyanide, 84, 85
Semiquinone, oxidation by ferricyanide, 84
Siderophores, 6
Silica gel, imidazole-functionalized, 42
Silica gel-supported porphyrins, 42
Sperm whale metmyoglobin, 7
Steapsin, 89

Strapped porphyrins, 21
Streptococcus faecalis, 212
Stryene/1-vinylimidazole copolymers, 43
Submitochondrial particles, 212
Substrate binding, 171-175
Superoxide, chemical properties, 212-215
  as oxidant, 213
  as reductant, 213
  definition, 211
  formation in biological systems, 212
  reactivity with organic substances in water, 215
  role in oxygen metabolism, 211
  in oxygen toxicity, 212
  systems forming detectable amounts of, in presence of oxygen, 214
  systems with no reactions, 210
  toxicity of, 218-220
    extrapolation from in vitro to in vivo conditions, 223, 225-227
    in vitro, 219-220, 222
Superoxide dismutases, 6, 211, 215-218
  biological distribution, 227
  definition, 215
  function of, 227
  metal ion catalysis of, 217, 218
  presence in anaerobes, 229
  properties of, 217
  protection against oxygen toxicity, 211
Superoxide-mediated Fenton chemistry, 220-223
Superoxo complexes, 49
  infrared spectra, 49
  resonance Raman spectra, 49
Synthetases, 6

Tailbase porphyrins, 20
Tailed picket fence porphyrins, 45, 46, 47
D-Tartaric acid, treatment with H_2O_2 and Fe^{2+}, 221
Temperature-jump experiments, 36
Terpolymers, 43
2,3,7,8-Tetrachlorodibenzo-*p*-dioxin, 78
2,2,6,6-Tetradeuterocyclohexanol, 109
Tetradeuteronorbornane, P-450 catalyzed hydroxylation of, 110
Exo,exo,exo,exo-Tetradeuteronorbornane, P-450-catalyzed hydroxylation of, 110
Tetrahydrobiopterine, as reducing agent, 140

INDEX

Tetrahydrofuran, 95
Tetrahydrothiophene, 95
Tetralin hydroperoxide, 114
1,2,4,5-Tetramethylimidazole, 61
Tetranitromethane, reduction of, 213
Tetra(p-chlorocarboxyphenyl)porphyrin, 44
Tetraphenyl porphyrins, 20
Tetra(p-methoxyphenyl)porphyrin, 29
Thiolate ligand, 103
Tiron, oxidation by superoxide, 213
Tissue oxidative phosphorylation, 9
α-Toxopherol, oxidation by superoxide, 213
Trace metal ions, nature of in in vitro degradative processes, 223, 224
Transferrin, 6
Transferrins, study of resonance Raman between dioxygenases and, 170
Transition metal ions, as reductants of superoxide, 215

Trypsin-solubilized reductase, 84
Tyrosine, 168

Udenfriend system, 140, 141, 142
Ullrich system, 142

Vanadium-oxo compounds, hydroxylation of aliphatic centers by, 138-139
1-Vinylimidazole, copolymer of, with styrene, 41
Vitamin B-12 cofactor, 6

Wacker cycle, 131

Zinc/copper protein, *see* Erythrocuprein
Zinc/copper superoxide dismutase, 213, 230
  reduction of, 213, 215
  in seeds, 229, 230